Coloniality and Decolonisation in the Nordic Region

This book advances critical discussions about what coloniality, decoloniality, and decolonisation mean and imply in the Nordic region.

It brings together analysis of complex realities from the perspectives of the Nordic peoples, a region that is often overlooked in current research, and explores the processes of decolonisation that are taking place in this region. The book offers a variety of perspectives that engage with issues such as Islamic feminism and the progressive left; racialisation and agency among Muslim youths; indigenising distance language education for Sami; extractivism and resistance among the Sami; the Nordic international development endeavour through education; Swedish TV reporting on Venezuela; creolizing subjectivities across Roma and non-Roma worlds and hierarchies; and the whitewashing and sanitisation of decoloniality in the Nordic region.

As such, this book extends much of the productive dialogue that has recently occurred internationally in decolonial thinking but also in the areas of critical race theory, whiteness studies, and postcolonial studies to concrete and critical problems in the Nordic region. This should make the book of considerable interest to scholars of history of ideas, anthropology, sociology, cultural studies, postcolonial studies, international development studies, legal sociology, and (intercultural) philosophy with an interest in coloniality and decolonial social change.

Adrián Groglopo has a PhD in sociology and is a senior lecturer at the department of social work at the University of Gothenburg, Sweden. His research focuses mainly on racism, the coloniality of knowledge, social movements, and north–south political, economic, and environmental relations. Groglopo has worked on several governmental projects regarding structural racism in Sweden, among others as Secretary of Enquiry at the Government's Enquiry on Structural Discrimination between 2004 and 2006. He acted as president of the Anti-racist Academy—an association that gathers around 60 researchers within the area of race and racism. In this context, Groglopo also led the production of a series of 17 filmed interviews with anti-racist researchers and activists in Sweden, available at www.antirasistiskaakademin.se. He also led the NOS-HS funded project *Decolonial critique, knowledge production and social change in the Nordic countries* (DENOR)—a series of research workshops that gathered around 200 researchers in the Nordic region. His latest publications include two co-authored articles (in Swedish) about coloniality and social work (2021), and a co-authored article with the comic artist Amalia Alvarez about racism and migrant representativity in comics (2022).

Julia Suárez-Krabbe is Associate Professor in Cultural Encounters at the Department of Communication and Arts, Roskilde University, Denmark, and Distinguished Research Associate at the Ali Mazrui Centre for Higher Education, University of Johannesburg, South Africa. Her work centres on racism, human rights, development, knowledge production, education, and decolonisation in Europe and the Americas. Her latest work includes the co-authorship of the report *Stop Killing Us Slowly. A Research Report on the Motivation Enhancement Measures and the Criminalization of Rejected Asylum Seekers in Denmark* from 2018, which includes examinations of state-sanctioned racism in Danish deportation camps and was written in close collaboration with the refugee movement in Denmark. Her work additionally revolves around the ontological, epistemological, and existential dimensions of decolonisation in Denmark. Julia is the author of *Race, Rights and Rebels. Alternatives to Human Rights and Development from the Global South* (2016).

Routledge Research on Decoloniality and New Postcolonialisms
Series Editor: Mark Jackson, Senior Lecturer in Postcolonial Geographies, School of Geographical Sciences, University of Bristol, UK.

Routledge Research on Decoloniality and New Postcolonialisms is a forum for original, critical research into the histories, legacies, and life-worlds of modern colonialism, postcolonialism, and contemporary coloniality. It analyses efforts to decolonise dominant and damaging forms of thinking and practice, and identifies, from around the world, diverse perspectives that encourage living and flourishing differently. Once the purview of a postcolonial studies informed by the cultural turn's important focus on identity, language, text, and representation, today's resurgent critiques of coloniality are also increasingly informed, across the humanities and social sciences, by a host of new influences and continuing insights for different futures: indigeneity, critical race theory, relational ecologies, critical semiotics, posthumanisms, ontology, affect, feminist standpoints, creative methodologies, post-development, critical pedagogies, intercultural activisms, place-based knowledges, and much else. The series welcomes a range of contributions from socially engaged intellectuals, theoretical scholars, empirical analysts, and critical practitioners whose work attends, and commits, to newly rigorous analyses of alternative proposals for understanding life and living well on our increasingly damaged earth.

This series is aimed at upper-level undergraduates, research students, and academics, appealing to scholars from a range of academic fields including human geography, sociology, politics, and broader interdisciplinary fields of social sciences, arts, and humanities.

Transdisciplinary Thinking from the Global South
Whose problems, whose solutions?
Edited by Juan Carlos Finck Carrales and Julia Suárez-Krabbe

The Coloniality of Modern Taste
A Critique of Gastronomic Thought
Zilkia Janer

Coloniality and Decolonisation in the Nordic Region
Edited by Adrián Groglopo and Julia Suárez-Krabbe

For more information about this series, please visit: https://www.routledge.com/Routledge-Research-on-Decoloniality-and-New-Postcolonialisms/book-series/RRNP

Coloniality and Decolonisation in the Nordic Region

Edited by Adrián Groglopo and
Julia Suárez-Krabbe

Routledge
Taylor & Francis Group
LONDON AND NEW YORK

First published 2023
by Routledge
4 Park Square, Milton Park, Abingdon, Oxon OX14 4RN

and by Routledge
605 Third Avenue, New York, NY 10158

Routledge is an imprint of the Taylor & Francis Group, an informa business

British Library Cataloguing-in-Publication Data
A catalogue record for this book is available from the British Library

ISBN: 978-1-032-27486-7 (hbk)
ISBN: 978-1-032-27567-3 (pbk)
ISBN: 978-1-003-29332-3 (ebk)

DOI: 10.4324/9781003293323

Typeset in Times New Roman
by SPi Technologies India Pvt Ltd (Straive)

To the memory of Gerard Gbeyo and all the victims of racism and white ignorance.

Contents

List of illustrations ix
Acknowledgements x
Notes on the contributors xii

Coloniality and decolonisation in the Nordic region: an introduction 1
ADRIÁN GROGLOPO AND JULIA SUÁREZ-KRABBE

1 **Surviving like Scheherazade. Veiled women and liberalism: the trap of the progressive left** 22
HOURIA BOUTELDJA

2 **Racialisation in a "raceless" nation: Muslims navigating Islamophobia in Denmark's everyday life** 37
AMANI HASSANI

3 **Enriching Sami language distance education** 51
HANNA HELANDER, SATU-MARJUT PIESKI, AND PIGGA KESKITALO

4 **The virtue of extraction and decolonial recollection in Gállok, Sápmi** 68
GEORGIA DE LEEUW

5 **Coloniality of knowledge and the responsibility to teach: Nordic educational interventions in the "South"** 89
JELENA VIĆENTIĆ

6 **Swedish television reporting on Venezuela as *damnation*** 110
JUAN VELÁSQUEZ ATEHORTÚA

7 Creolizing subjectivities and relationalities within Roma-gadje research collaborations 126
IOANA ȚÎȘTEA AND GABRIELA BĂNCUȚĂ

8 Decoloniality: between a travelling concept and a relational onto-epistemic political stance 145
MADINA TLOSTANOVA

Index 164

Illustrations

Figures

3.1 Sunna Kitti, the Pilot project on distance education in Sámi languages 59
4.1 Den Svenska Gruvan 2021 (own website stillframe) 79
4.2 Preparations for Tjáhppis Rájddo manifestation during the 2014 Jokkmokk winter market (Photo: Tor Lennart Tuorda) 80
4.3 Profile picture and community background image of the *What local people* Facebook community page (stillframe accessed March 3, 2022) 81
4.4 Anders Sunna's art installation, used as the focal point in Sofia Jannok's music video *We are still here* in 2016 in Gállok (anderssunna.com) 82
7.1 My mother makes the best polenta in the world (Photo taken by Gabriela with her phone) 134

Maps

3.1 Map of Sami region (University of Lapland) 52
4.1 Sápmi: The Sami Homelands, Jordan Engel (The Decolonial Atlas n.d.) 71

Acknowledgements

This book is the outcome of a research network project funded by NOS-HS (Joint Committee for Nordic Research Councils in the Humanities and Social Sciences). The research network *Decolonial critique, knowledge production and social change in the Nordic countries* (DENOR) gathered over 200 researchers and activists in the Nordic countries in a series of conferences and workshops organised by scholars from Denmark, Sweden, Norway, and Finland between 2018 and 2019. The aim of these events was to address coloniality's social, political, epistemic, and ontological production of absences and their connection to contemporary problems connected to knowledge and education in the Nordic countries. It examined how these may reproduce or be complicit in the problems of racism, and social and political exclusion we are facing in the region.

Our ambition with the network was to create a generative environment for researchers who study racialisation, social exclusion, and alternatives to the existing social order from the perspective of its margins. We also aimed to critically adapt decolonial critiques to understand these issues. The workshop series was intended to facilitate the building of networks across the Nordic region on decolonial critique, knowledge production, and social change in these countries, which would continue to build innovative interdisciplinary work across the fields of studies of racialisation and racism, indigenous studies, and migration studies. The workshops were organised in three cities and universities: the University of Gothenburg in Sweden, the Norwegian University of Science and Technology in the city of Trondheim, Norway, and the University of Helsinki in Finland. Each conference was organised around a theme that framed the conference discussions and paper presentations as well as the invited keynote speakers. The first, held in Gothenburg, was about theoretical frameworks for coloniality; the second, in Trondheim, was about decolonial learning and pedagogy; and the third, in Helsinki, was organised around thinking on decolonial and anti-racist activism from the framework of decolonial critiques. Besides the two authors of this introduction, two other scholars were involved in this project—Professor Suvi Keskinen (University of Helsinki) and Associate Professor Stine Bang Svendsen (Norwegian University of Science and Technology). To Suvi and

Stine: thank you for all your work and support in the making of the decolonial network.

We want to thank all the participants of the three conferences and all the workshops that were organised during the period. Many people have been involved in making the conferences and workshops possible. First, a recognition to NOS-HS for having financed the series and made possible the construction of the decolonial platform in the Nordic countries, and also to the department of Social Work at the University of Gothenburg for administrating the finances and logistics.

We especially want to acknowledge and thank all the keynote speakers who travelled from different locations around the world to contribute with their knowledge and experiences, and who enriched all the participants with thoughtful and spiritual inspiration. Their input was pivotal to the work of opening further paths for decolonised sciences and knowing in this part of the world: May-Britt Öhman Tuohea Rim, Vanessa Thompson, Nelson Maldonado-Torres, Pigga Keskitalo, Rosalba Icaza, Deise Nunez, Lesley-Ann Brown, Nokuthula Hlabangane, Alfian Bin Sa'at, and Houria Bouteldja.

We also want to warmly thank all the panellists who contributed with important insights about racism and decolonial thought: Maimuna Abdullahi, Amanj Aziz, Faith Mkwesha, Eóin Cuinneagáin, Liam Lillis Ó Laoire, Arwa Awan, Oda-Kange Midtvåge Diallo, Thulile Gamedze, Anja Márjá Nysø Keskitalo, Ellen Marie Jensen, and Peder Brede Jensen. Many others have also been involved in the logistics and administrative organisation, as for example Maximilian Weik.

The network provided the structure needed for building a decolonial research community in the Nordic countries and enabled Nordic researchers to take part in this internationally growing field of study. The initiative also facilitated the inclusion of early career researchers to this field and provided stakeholders in the field with the opportunity to provide valuable input.

Notes on the contributors

Gabriela Băncuță lives in Helsinki, Finland. She is currently attending a Finnish language and literacy course.

Houria Bouteldja is a founding member of le Parti des Indigènes de la République, a decolonial political organisation based in France. She has written numerous theoretical strategic articles on decolonial feminism, racism, autonomy, and political alliances as well as articles on Zionism and state philosemitism. She is the author with Sadri Khiari of *Nous sommes les indigènes de la république* (Editions Amsterdam) and *Whites, Jews and Us, towards a politics of revolutionary love* (Semiotext(e)). She recently resigned from the party, but she is still a decolonial activist.

Amani Hassani is a Leverhulme postdoctoral fellow in Sociology at Brunel University in the UK. Her research looks at processes of racialisation, spatialisation, and urban life among Muslims in the Global North.

Hanna Helander is a researcher at the University of Lapland in the *Socially Innovative Interventions to Foster and Advance Young Children's Inclusion and Agency in Society through Voice and Story* (ADVOST) research project. She is also a project manager for the *Pilot Project on Distance Education in the Sámi Languages* and PhD scholar at the University of Oulu.

Pigga Keskitalo has recently joined the University of Lapland as a professor. She earned her PhD in Education in 2010. She is also an adjunct professor at the University of Helsinki. Currently, she is working on the ADVOST research project, which deals with Sami language distance education.

Georgia de Leeuw is a Political Science PhD candidate at Lund University, Sweden, focusing on extractivism.

Satu-Marjut Pieski worked as a Sami language teacher in the *Pilot Project on Distance Education in the Sámi Languages*. She is currently working as a senior inspector at the Regional State Administrative Agency.

Ioana Țîștea is a PhD researcher at Tampere University, Faculty of Education and Culture. She is currently finalising her doctoral thesis, entitled

Creolizing the epistemologies and methodologies of Finnish migration research: Autoethnographic stories and theoretical entanglements.

Madina Tlostanova is professor of postcolonial feminisms at Linköping University. Her most recent books include *What Does it Mean to be Post-Soviet? Decolonial Art from the Ruins of the Soviet Empire* (2018), *A new Political Imagination, Making the Case* (co-authored with Tony Fry, 2020), and the forthcoming *Narratives of Unsettlement.*

Juan Velásquez Atehortúa is associate professor in gender studies at the University of Gothenburg. His research interests revolve around barrio feminism and decolonial thinking from Scandinavia and Latin America and the Caribbean regions. He has published his results in anthologies and peer review journals in English, Spanish, and Swedish.

Jelena Vićentić holds a PhD from the University of Belgrade, Serbia. Her work focuses on decolonial approaches to understanding development. Her recent publications include *Europe Facing Its Colonial Past* (co-editor and contributor, 2021) and *When Scanguilt meets reality show: doing-good entertainment and media reproduction of colonial imaginary of the 'South'* (2021).

Coloniality and decolonisation in the Nordic region

An introduction

Adrián Groglopo and Julia Suárez-Krabbe

This book presents a selection of texts that, through their engagement with decolonial and other Global Southern perspectives, provide original readings of some of the problems faced in the Nordic region.[1] The contributions included in this volume by no means do justice to the vastness, complexity, and diversity that comprise the Nordics, and a lack of perspectives from former and contemporary Danish colonies represents a significant absence. Although such perspectives can be found in recent publications (Tafari-Ama 2020; Christensen and Heinrich 2014; Graugaard 2014; Naum and Nordin 2013), there is still a long way to go in establishing fruitful conversations about the meaning and implications of coloniality, decoloniality, and decolonisation, not only in Denmark but also in the Nordic context more broadly. Such a collective endeavour cannot bypass the perspectives and historical experience of people in the Nordic region who are characterised as non-belonging, absent, criminal, and/or barbaric in general, including 'non-Western' migrants and refugees, Afro-Nordics, and Muslim communities, as well as the Romani and the Indigenous communities of the region such as the Sami and Inuit. That is precisely where this book comes in. We are interested in a serious approach to the question of how coloniality—as a specific colonial matrix or pattern of power grounded in the idea of race and racism as an organising principle of the modern/colonial capitalist world—permeates not only the social, political, economic, and environmental systems of this region but also social relations in areas such as labour, nature, sex/gender and sexuality, subjectivity and authority, and the very process of knowing and knowledge production (Quijano 2000).

Although the volume does not cover all the complexities, peoples, and localities in the Nordic region, it does provide important contributions to decolonial critiques. Many of us have been academically engaged for many years while also being involved as activists in different spaces of decolonial and anti-racist resistance. In these spaces, Indigenous, anti-racist, decolonial feminist, and other Global Southern perspectives have been key and nurtured our analyses of the problems in the region in their connection to racial capitalism and the geopolitics of imperialism. In this sense, the volume also engages with and learns from the struggles and problems faced by people engaged in anticolonial, antiimperialist, decolonial, and anti-racist struggles

DOI: 10.4324/9781003293323-1

in the European region and around the world. This means that our knowledge has been shaped and inspired by different critical intellectual environments from Latin America, the Caribbean, and Africa, as well as from the black radical tradition, black Marxism, Indigenous scholarship, and critical Muslim studies.

To varying degrees, we have also been formed in Nordic universities, meaning that we have drunk from the fountain of Eurocentric scientific knowledge and engaged in discussions in the growing areas of study in the Nordic region on race, whiteness, critical theory, and postcolonial theory. These latter critical perspectives have provided several important insights into the characteristics of race relations and racism in the political and social spectrum of the Nordic countries that were rarely addressed before the beginning of the 1990s (Knocke 1987; Andersen and Nielsen 1989; Ålund and Schierup 1991; Jonsson 1995; De los Reyes, Molina, and Mulinari 2002; Gullestad 2004; Loftsdóttir and Jensen 2012; Hübinette and Lundström 2014; McEachrane 2014; Hervik 2015, 2019; Groglopo 2015; Vitus and Andreassen 2015; Jensen 2018; Padovan-Özdemir and Øland 2022). Through a focus on questions of identity, representation, discourse, affectivity, and power, such works have provided important contributions to understanding the cultural and symbolic dimensions of the Nordic region. However, there is a need for further research that includes in its analysis racial capitalism and imperialist politics, including the racial division of labour in the region. Additionally, while many scholars adhere to the critiques of Eurocentrism, few reflect a dedicated engagement with theories and perspectives beyond the Eurocentric paradigm(s). They have also evaded addressing the coloniality of knowledge, including serious engagement with the positivist assumptions and basic tenets of methods and methodologies that have been addressed by countless Global Southern scholars (Fanon 1967; Wa Thiong'o 1986; Fals Borda 1979, 1981; D'Amico-Samuels 1997; Tuhiwai-Smith 1999; Sandoval 2000; Vasco 2002, 2007; Gordon 2006a, b, 2011; Rivera Cusicanqui 2012).[2] As Rosalba Icaza and Rolando Vázquez argue in relation to decolonising the university:

> the task [...] is not an ideological position but an epistemic stance that struggles against the ignorance of monocultural approaches. Under this analysis Eurocentrism is detached from being an issue of identity to become an epistemic problem, namely the problem of monocultural approach to knowledge practices, to research, to teaching and learning.
> (Icaza and Vázquez 2018, 116)

The chapters in this volume all work against this ignorance of monocultural approaches, offering critical purviews to a diversity of issues. Houria Bouteldja's contribution offers an illuminating decolonial critique of Islamic feminism and its investment in a colonial, white left, and Western feminism through the analysis of an instance of voluntary unveiling in France. This is the only chapter in the volume that does not engage specifically with the Nordic region. However, we have included it here because her analysis also provides critical and often censured nuances against the logic and political workings of Islamophobia in the Nordic region. Amani Hassani's chapter highlights the specificity of Islamophobia in Denmark by zooming in on racialisation, agency, and everyday resistance among Muslim youths. Importantly, Hassani's approach reverses the otherising and racialising gaze predominant in academia, thereby exposing how Muslim youths adopt different strategies to navigate the widespread and normalised Islamophobia in Denmark.

In the case of Indigenous populations, and specifically the Sami, Hanna Helander, Satu-Marjut Pieski, and Pigga Keskitalo's chapter addresses the challenges faced by Sami language educators and the importance of indigenising distance language education for Sami children as a process of cultural and epistemic resistance and survival. Bridging traditional Sami child-rearing practices and institutionalised learning, the chapter contributes to pedagogies that are inclusive and representative of Indigenous peoples. Following this, Georgia de Leeuw's chapter focuses on extractivism in Gállok in the north of Sweden, arguing that access to Indigenous land is managed on the basis of "coloniality of access". The chapter also focuses on successful narrative processes from Sami resistance in Gállok that have effectively postponed a government decision to grant a mining permit for nearly a decade and contributes to informing other processes of Indigenous and local resistance to extractivism.

Writing on the coloniality of knowledge and "the responsibility to teach", Jelena Vićentić's chapter explores educational programmes supported, sponsored, and promoted by Nordic actors in their endeavours of international development in the "South". It reveals how such programmes operate with narratives and methodologies that perpetuate the coloniality of power and knowledge, including systematic erasures of epistemologies that challenge the existing hegemonic order. Thereby, the chapter unpacks one of the arenas in which the Nordic region is decisively invested in maintaining a colonial, capitalist, and imperial world order. In his chapter on Swedish TV's reporting on Venezuela, Juan Velásquez Atehortúa trenchantly unpacks the specific ways in which such a world order is maintained through a pedagogy of cruelty (Segato 2016) that trains its audience "to tolerate acts of cruelty against a substantial part of the Venezuelan population, which is produced as dispensable through the very same reporting." (this volume, page 111).

The last two chapters can be seen as key contributions beyond (de)coloniality. Ioana Țîștea and Gabriela Băncuță bravely engage in a process of creolizing subjectivities across Roma and non-Roma worlds and hierarchies. Through a dialogical and co-authorial methodological approach, they work against and beyond hierarchical divisions through an ongoing, open-ended process of becoming based on mutual self-reflection and self-aware experimentation. The chapter is an important contribution to decolonising methodologies, knowledge, and subjectivities through its contestation and re/negotiation of boundaries and hierarchies. Lastly, Madina Tlostanova offers an incisive critique of the whitewashing and sanitisation of decoloniality in the Nordic region, arguing that current global challenges go beyond the original decolonial focus on the intersection of race and capitalism and include climate change, chronophobia, defuturing, and global unsettlement. The chapter makes an important contribution to decolonial scholarship in the region, challenging it to do better and go beyond current thought to reach a better understanding of the decolonial potential in the future and its applicability in other places such as Nordic Europe. Tlostanova's reading of decoloniality is highly important; it converges with our own analyses of its adoption in the Nordic region in some regards, while differing significantly on others about what may or may not constitute a decolonial project. This introduction is not the place to enter into such otherwise important discussions, however in it we present our take on many of the issues that Tlostanova raises. One of these is the poststructuralisation of decoloniality that we address below.

The postcolonial and the decolonial

The decolonial perspective, and discussions regarding coloniality and decolonisation in general, emerged in Nordic—and European—academia through several different but interrelated streams that we present briefly in the following. It is important to note that this presentation does not follow a hierarchical or chronological order, but rather tries to make visible the different processes and movements that converged and often nurtured each other. Firstly, it appeared in the work of young scholars and activists from Latin American backgrounds—among whom we, the editors of this book, found ourselves—around the second half of the 2000s.[3] We began to establish coalitions in the margins of academia with other racialised populations and movements in the Nordic countries and the rest of Europe; during this time the role of the Decoloniality Europe network was important for many of us. Secondly, it was driven by the hard work of the Indigenous populations in the Nordic region, notably the Sami and the Greenlandic Inuit, who were engaged in discussions on decolonisation with other—mainly English-speaking—Indigenous scholars and activists around the world. Thirdly, it was the influence of the descendants of those enslaved during the Danish involvement in transatlantic trade from former Danish colonies. Notably, this stream emerged forcefully through art and includes the work of scholars, artists, writers, and activists such as La Vaughn Belle, Jeanette Ehlers, Bright Bimpong, Tami Navarro, Imani Tafari-Ama, Tiphanie Yanique, and Shelley Moorhead (see for example Moorhead 2017). The continued anti-racist work of Afro-Nordic associations and movements have been key for central discussions and actions—as was the Black Lives Matter demonstrations in 2020 in main cities of the Nordic countries—against antiblack racism. Equally important have been the mobilisations against the deportation regime and the increasing implementation of racism through legal measures in the region. Fourthly, it was the construction of anti-racist movements—pointing to structural racism as the main structure of intersectional domination in the Nordic countries—that emerged out of cooperation and organisation between scholars and activists, for example the Anti-Racist Academy in Sweden and Marronage in Denmark. Finally, in response to the global imperial doctrine of "the war on terror," was the mobilisation of critical Muslim youths to condemn the political, mediatic, and organisatoric rise of racism against Muslims, namely Islamophobia. These important developments in the long history of anti-racist and/or anticolonial resistance among the highly diverse populations of the Nordic region are all too often rendered non-existent in Nordic academia (Keskinen, Skaptadóttir and Toivanen 2019), which instead tends to see the decolonial as a sort of spin-off from the postcolonial (see also Tlostanova, this volume).

Lewis Gordon (2021) addresses these problems in depth in his discussion of the relationship between decoloniality and poststructuralism. Influential to the Euromodern academy from the late 1960s through to the early 2000s, poststructuralism was seen by many as a liberatory theoretical development that nonetheless came into crisis as its compatibility with neoliberalism, imperialism, and colonialisation became clear. In the face of such problems, most poststructuralists resorted to "the age-old technique of rebranding" (Gordon 2021, 15), presenting themselves through critical theory and decoloniality. Following Gordon, this rebranding is an acute problem given the undertheorised nature of decoloniality, and indeed it works as a form of coloniality within decoloniality, as evident in prolific, but largely inconsequential, statements on "positionality"—interventions often moral in nature presented as political. In the end, structural inequality remains intact, and the "privileged" successfully evade responsibility by posing moralistic solutions to political and structural problems (Gordon 2021, 15–16; see also Maldonado-Torres 2020; Suárez-Krabbe 2016). A similar problem appears with the concept of "colonial complicity" (Vuorela 2009; Keskinen et al. 2009), used by some scholars to refer to the Nordic region's involvement in colonial and imperial projects (see also Tlostanova, this volume).[4] It is important here to stress that the problem of taking a poststructuralist approach to understanding coloniality is that it is also a way for Eurocentric critical theory to whitewash and depoliticise the radical concepts and projects emanating from the colonial subjects in the Nordic region and the colonised regions of the world. This mirrors the "undertheoretisation" that Gordon (2021) problematises and reduces the political to the moral. This is why it is critical that we reclaim the political, which includes understanding—and politically organising against—the materiality of colonial politics in contemporary Nordic societies.

In addition to the above, further differences between the postcolonial and the decolonial exist. These are, following Ramón Grosfoguel (2011), genealogical, epistemological, and political. The genealogical difference concerns the approach to modernity and coloniality. The decolonial perspective fundamentally challenges the myth of modernity (Dussel 1993) by underscoring that coloniality is its other side (Mignolo 2011). What Dussel (1993) called the myth of modernity consists of an understanding of modernity as an intra-European phenomenon, whose central axis is the industrial revolution.

This myth justifies the fundamental falsehoods of conflating Europe with modernity, by which Europe is understood as the cradle of rights, freedom and equality, knowledge, and technology, and even as the heart of human development. Such an understanding is also based on the idea of a largely homogeneous (white) Europe that erases the histories and perspectives of many of its populations (Suárez-Krabbe 2014) and presents the rest of the world as the stage for the spread of Euromodernity that will bring "prosperity" and advancement to all people. In the decolonial perspective, the year 1492 when Columbus arrived to *Abya Yala* (the Americas) is fundamental to the emergence of Euromodernity and coloniality. Taking into account this historical period allows a deeper understanding of the legacies of colonialism and their imbrication with racial capitalism (Cox 1959; Robinson 1983; Montañez Pico 2020), breaking the "monopoly" that Euromodernity has on freedom, emancipation, and liberation. It allows displaying the fundamental role that past anticolonial struggles—the most famous, perhaps, being the Haitian revolution in 1791—played in global emancipatory ideas and practices subsequently co-opted by the European powers in favour of defending their (white) supremacy (James 1938; Robinson 1983). As Grosfoguel underlines, it is in this period between 1492 and the mid-19th century that Europe gains the economic and military power necessary to construct itself as politically and culturally superior through the production of specific (colonial/racist/sexist) knowledge to be seen, processed, and taught as universals (see also Williams 1944; Rodney 1972; Davis 1981; Blaut 1993; Tesfahuney 2005). Clouding the importance of coloniality allows Euromodernity to conceal its foundation in racist and white supremacist ideology and to deny its continued investment in colonial capitalist expansion, inferiorisation, and exploitation of nature, labour, and spaces.

The political distinction between the postcolonial and the decolonial derives from the above genealogical distinction, as our understanding of modernity will decisively shape the political projects to which we aspire (Grosfoguel 2011). If we take the myth of modernity for granted and understand it as an emancipatory project, and not as a conquering/colonising/civilisatory one, then our political goal is to reform modernity from its epistemological interior, that is from within Eurocentric knowledge, its bourgeoisie logics, imperialist silences and horizons of political (im)possibilities. This has been the goal of postmodernism, poststructuralism, and other "*posts*" … (Meiksins Wood 1986; Dirlik 1994; Anderson 1998; Kohan 2022). This point leads us to the epistemological distinction. As we have discussed previously, postcolonial thought is also grounded in the poststructural approach, where the most known aim is deconstruction (of discourses, of the subject …), favouring fragmentation and difference in times of neoliberal and imperialist advances, as well as in contexts of parliamentary advances of neo-fascist and neo-conservative political parties in Europe and the Nordic region. Such an aim that does not politically dismantle the colonial matrix of power, more often than not continues to reproduce white supremacy, racial capitalism, and imperialism. Additionally, in the Westernised universities, we are mainly presented with the thinking of a small number of men (also revealing a sexist axis) on the problems of the world and their solutions, neglecting a great number of those who throughout time have produced knowledge about, against, and beyond the colonial capitalist system (see Abdel-Malek 1963; Davis 1981; Reddock 1994; Achcar 2008; Grosfoguel, 2013; Khotari et al. 2019; Fink-Carrales and Suárez-Krabbe 2022). Therefore, the pledge of decolonial, anticolonial, and other Global Southern perspectives is for epistemic, or cognitive, justice (Santos 2014)—for the inclusion of thinkers from places, spaces, and genders that have been obscured or wiped out by modernity/coloniality (see also Oyěwùmí 1997; Lugones 2010). The aim is not to abolish the knowledge of those few European men but to diversify and open up opportunities for other epistemologies, humanities, social and natural sciences, and cosmologies (see also Khatibi 1981; Wallerstein 1996; Santos 2014; Mignolo and Walsh 2018). However, diversification in and by itself can—and easily does—fall into the traps of the neoliberal project when decoloniality is understood merely in terms of epistemic diversification. When this is so, decoloniality changes the players but not the game (Gordon 2021, 15). Decoloniality is political to the extent it does not become an end in itself but is instead, as Catherine Walsh (2018) argues, a decoloniality *for*, not simply a decoloniality *from*. As Gordon has noted (Gordon 2021, 16), "the aim of decolonization *from* leads to no one ever being decolonized enough except, perhaps, the one who poses the problem of decolonization, with decoloniality being its purest commitment." Decoloniality *for*, on the other hand, requires political thinking and organised action beyond decoloniality.

Coloniality and decolonisation in the Nordic region

We have argued that much of the decolonial scholarship in the Nordic region approaches coloniality through a poststructuralist and postcolonial lens. We have also underlined how such scholarship reproduces the coloniality of knowledge (Grosfoguel 2013; Suárez-Krabbe 2012) by ignoring and/or depoliticising the radical political concepts and projects emanating from the colonial subjects in the Nordic region and the colonised regions of the world resulting in "undertheoretisation" (Gordon 2021). Our hope with this volume is to nurture scholarship in the region that engages in reclaiming the political, which involves theorising the materiality of colonial politics in contemporary Nordic societies. In this final section, we lay out some key issues in this regard.

The Nordic region has a reputation that attracts the interest and attention of many progressive scholars and politicians around the world, a reputation that is also an irrefutable "truth" to many people in the region—a region with minimal class inequalities, where all work is dignified and where human rights are respected and race and gender inequalities absent, a region that has built strong welfare states based on tolerance, equality, and solidarity and is thereby a model to be followed by those seeking equality and democracy (see Habel 2012; Naum and Nordin 2013; Lauesen 2021). There is also a prevailing idea that the Nordics played only a minor role in the imperialist wars and expansionism of European colonialism, and that consequently race and racism are "inventions" taken from the US and adopted uncritically by some scholars and activists. Such falsehoods sweep under the carpet the long and complex ties that the Nordic region has had historically with the emergence of European colonial expansion and its continued investment in perpetuating a globalised racist, capitalist, and patriarchal system. These complex links are being revealed by scholars in the Nordic region who continue to explore Nordic (particularly Danish and Swedish) overseas (neo)colonialism in Latin America, Asia, and Africa. Such research includes the political and economic interests of these colonialisms (Höglund and Burnett 2019, 3; Lauesen 2021), also in the Nordic region itself, and point to "the fact that Sámi land was colonized by Denmark-Norway, Sweden, Finland, and Russia over a long period of time, and that the region remains, to all effects and purposes, a colony of these nation states" (Höglund and Burnett 2019, 2).

In spite of such scholarly attention, even some progressive scholars place the Nordic region in a position that is surprisingly coherent with the nation-branding narratives of the Nordic countries—tolerant, peaceful, and humane welfare states that have been peripheral to the European imperial endeavour, where equality and justice reign above racism. According to Marta Padovan-Özdemir and Trine Øland (2022, 43), "the current Nordic self-perception as 'progressive humanitarians' was originally powered by Swedish Gunnar and Alva Myrdal's 20th-century defining ideas about social engineering as the means to realise the liberal principles of freedom, equality and justice" are inherently racist. Indeed, racism in the Nordic countries is tied to the liberal/racial foundations of capital accumulation and uses the welfare state and its practices of differentiation, assimilation and exclusion (Ålund and Schierup 1991; Goldberg 2002; Schierup, Hansen, and Castles 2006; Sawyer and Habel 2014; Keskinen 2016; Small 2018; Padovan-Özdemir and Øland 2022; Mulinari and Neergaard 2022). These policies and practices of course have real effects on those of us who are here as colonial subjects. Often designated as non-white/non-Western, our everyday lives, as well as those of our nearest and dearest, are characterised by having to deal with processes of racialisation and otherisation—not seldom discrimination and violence—where we face exclusion, police profiling, criminalisation and face greater consequences for crimes, including deportation and health problems (Hassani, this volume; Ţîştea and Băncuţă, this volume; Groglopo and Ahlberg 2006; Greve 2016; Myong and Bissenbakker 2016; Arce and Suárez-Krabbe 2018; Suárez-Krabbe and Lindberg 2019; Abdelhady, Gren, and Joormann 2020; Mulinari and Keskinen, 2020; Kamali 2021; Wallace et al. 2021). The colonial subject in the Nordic region may be targeted by such a racial hierarchy, but, if accepted as legal, we are granted specific but limited "privileges", such as a passport or working permit, a credit card, access to health care, and so on. Such inclusion is necessary inasmuch as it facilitates our integration in the economic and social system of racial capitalism, organised and upheld by white supremacy in the interest of capital accumulation in the region (Padovan-Özdemir and Øland 2022). In this sense, migrants from the South and Indigenous communities in the region can be seen as mere economic categories (see Cox 1959) whose (de)value in the capitalist labour market is determined according to the racial rules of white supremacy.

White supremacy places whiteness above ethics, morality, law, and politics (Gordon 2022, 110; Echeverria 2010, 61; Beliso-De Jesús and Pierre 2020) and grants licence to do harm without accountability to those who defend and obey its doctrine (Gordon 2022, 107). However,

> Getting rid of white supremacy doesn't entail the eradication of racism. While white supremacy is the thesis that whites must be superior and have everything [...] antiblack racism is the conviction that blacks must have nothing. Rejecting both involves understanding that all people deserve something, and that an all-or-nothing mentality is a false dilemma.
>
> (Gordon 2022, 165)

The racial colonial categories of white, black, and non-white are socio-politically constituted through the long history of Euromodernity and its political economy, with blackness expected to be nothingness, not-human, and indeed devoid of black consciousness (Gordon 2022), where other non-whites are a sign of mistrust and suspicion. As we have mentioned earlier, the modern capitalist economy that dominates globally, is racially organised (Robinson 1983, see also Fanon 1963; Nkrumah 1970; Hall 1980; Grosfoguel 2018). "White is a metaphor for power" wrote James Baldwin in 1963 (Baldwin 2017, 107), and such power continues today through Western imperialism and neocolonial relations that reproduce the centre-periphery pattern in which humans are economically, politically, and epistemically categorised hierarchically through the idea of race, a category constitutive to modern global capitalist expansion by colonial designs (Groglopo 2012). These categories work as classificatory systems according to phenotypic and cultural markers, but they are also epistemological positions of enunciation in these power structures. The locus of enunciation, as Walter Mignolo (2000) argues, is the position from which we think, talk, and act in the corpo-political and geopolitical spaces divided by the modern/colonial logics of domination and subjugation. As such, these relations of domination interpellate the colonial subject in different forms, sometimes epistemically in contradiction to what is expected from them (Grosfoguel 2018). Scholars from around the world have referred to different dimensions of this, among others as "colonization of the mind" (Wa Thiong'o), "spiritual brain-drain" (Fals Borda), "black skins, white masks" (Fanon) and "bad faith" (Gordon). What is at stake here is a critique of ideas of racial purity (Martínez and Davis 1994; Lugones 2003; Monahan 2011) on the one hand, and the question of political existence, that is to participate in setting the terms of the social structures and political horizons we envision (see, among many others, Aatar 2021 and Gordon 2022). Such participation can arguably be achieved within the current political system, but only if the racialised subject uses a "white mask" which, as Fanon showed, is tantamount to non-existence. In the Nordic region, for instance, Muslims can only enter the party-political field if they leave out their Muslimness, meaning if they pretend to be what they are not.

How the Nordic societies are structured and how the political and economic systems that organised them through modernity are reproduced is, then, also part of the coloniality of power. Indeed, the social control and domination of one core group over another is also a feature of modern capitalist societies, dividing the labour force using racial markers and policing such types of societal organisation. As Houria Bouteldja and Sadri Khiari (2021) argue, in this system of control and domination, subjects are shaped to fit the roles expected of their spaces, to become integrated, with a number of different policies applied to guide and control this part of the population (see also Tesfahuney and Grip 2007; Padovan-Özdemir and Øland 2022). On the other hand, the white majority is not targeted as non-integrated nor by policies of assimilation or integration. This is important because the expectations of authorities, the political discourse, and policies, and the Manichean narration of the media, shape the conditions and outcomes of social and political identities, constructing social opportunities (e.g., defining the political) and narrowing some sectors of society to certain conditions of living and future outcomes (Romero-Losacco 2018; Hervik 2019). However, if the targeted groups, in this case so-called non-Western migrants and their children as well as other racialised communities, accept social and political subordination, opportunities open up for them—as individuals, not as communities. In this social and political setting in the Nordic region, migrants—and increasingly children with migrant backgrounds (both those born in the Nordic countries to non-Western-born parents and those born outside the Nordic countries to non-Western-born parents) —are targeted by policies that put on them different controls to the rest of their Nordic counterparts (Khawaja 2015).

If there are political features that mark the Nordic countries, they are gender politics and feminist movements. These movements and their politics have been successful in challenging the gender structures that organise societies amongst men and women, opening spaces for other genders/sexes. They have, however also been criticised for basing their understanding of equality within the Nordic nation states on the assumption that they are racially homogenous, putting gender equality policies and discourses at the favour of capitalist, racist, and Islamophobic structures on a national and international level, including their neocolonial foreign politics (see Hassani, Vićentić, and Bouteldja in this volume; Bannerji 1995; Espinosa Miñoso 2009; Adlbi Sibai 2016; Zakaria 2021). Recognising that these politics and theoretical approaches of white feminism are constructed from the experiences and ethos of white bourgeoise women that are extrapolated as universal gender laws, some Nordic feminists have engaged in attempts to "decolonise" and deconstruct white feminist epistemologies and activism (Andersen, Hvenegård-Lassen, and Knobblock 2015; Andreassen and

Myong 2017; Keskinen, Stoltz, and Mulinari 2021). Such efforts notwithstanding, feminism continues to be a prison to many non-western and non-white people in Europe (Adlbi Sibai 2016) and precludes critical interventions from migrant and non-white feminists in the Nordics. In this context, intersectionality has become a concept commonly used in Nordic critical scholarship but, more often than not, as an assimilation into white liberal epistemologies—ideologically emphasising the (Westernized universal) individual subject as the point of departure of the analysis—and detached from its original political intent to address the imbricated ways in which processes of domination target people and communities according to the historical complex amalgamation of race, class, gender, and other situated conditions and categories (Crenshaw 1991; De los Reyes, Molina, and Mulinari 2002; Collins and Bilge 2016). Such an original intersectional approach also seeks to build political and epistemological bridges between different groups of people targeted by racism, sexism, heteronormativity, capitalism, and imperialism (Lugones 2003; Dahl 2021; De los Reyes and Mulinari 2020; Velásquez Atehortúa 2020). While the intersectional critical interventions in the Nordic region have meant the opening of white spaces, such as academia, for academic and research labour for migrants, Samis and non-white women, it is fair to say that these white feminist institutions and policies have embraced non-white and migrant women researchers as long as the white hegemony and its institutional settings are not challenged by them. In these white settings they risk becoming "colourful" exceptions who, through a greater amount of effort than their white peers, can achieve institutional recognition in Nordic academia (see Ahmed 2011; Vergès 2021).

To engage in analyses of capitalist hegemony and imperialist politics, and to take seriously the (re)production of race and racism in the Nordic region (as well as in the rest of Europe), is not only a theoretical necessity but also a political imperative. The fruits of neoliberal politics are evident in the increasing gap between rich and poor and its correlation with processes of racialisation and spatial racial segregation (Andersson and Molina 2018), as well as the dismantling and privatisation of some welfare state institutions (Ålund, Schierup, and Neergaard 2017). Indeed, the fragmentation and racial segmentation of the social is part of the colonial project that is still in process and obeys the current dominant expression of global racial capitalism, namely imperialism and its neoliberal project: modernity's latest civilisatory project (Dussel 1994; Karatani 2018). The analytical purview upon the Nordic region cannot overlook the Nordic partnership with the North Atlantic Treaty Organization (NATO) and with other secret services organised under the US-led war on terror on Muslim communities, nor the region's participation in imperialist wars like those in Afghanistan and Libya, and even neocolonial military interventions in some African countries, like Mali and Somalia. These imperial projects are deeply dependent upon the naturalisation of Islamophobia internally in the region, and aid the white supremacist interests in different national contexts. The Nordic region's investment in the NATO-Russia war in Ukraine (Amadae and Craig 2022; Mearsheimer 2022) also deserves attention and critical analysis. These are some of the reasons why it is important to reconsider the ways in which certain kinds of geopolitical interests and local logics of domination work in the region. How has the Nordic countries' insertion into global capitalism and imperial politics informed race and racialisation within the region and in its global expansion for capital accumulation? How has colonial/imperialist thinking in education, economics, and politics shaped the formation of class power in this region? How are race, class, and gender relations, as well as their forms of resistance, being re-articulated to obey capitalist interests and imperial politics? In what way does colonial/imperialist thinking serve the interest of class power and reinforce the hegemony of whiteness in the region? How does colonial/imperial thinking shape the Nordic countries' political discourses, media narratives, and policies? Questions like these are pivotal to the continued work of decolonisation in the Nordic countries.

As we mentioned earlier in the discussion on postcolonialism and decoloniality, decoloniality entails decolonising current structures for another world (Walsh 2018). While there are many perspectives presented in such discussions, Indigenous scholars and movements in the Americas and beyond have forwarded such critiques and projects since the inception of colonialism. Rafael Bautista's (2022) presentation of the decolonisation process resonates with these—as well as with our own—understandings of decolonisation, and is useful to detail in this final part of the introduction. Bautista notes that the decolonisation process has two moments, a negative and a positive. The negative moment is about systematically dismantling the narratives imposed by modernity. He names specifically three meta-narratives. The first is the capitalist narrative, an empirical narrative in front of us. According to this narrative, we as subjects are subsumed into the capitalist system, accepting its domination and exploitation of humanity and the environment, because "there is no alternative," "the system drives development and some sacrifices must be made to facilitate this," "it is the best system because it gives individuals freedom," etc. Behind this narrative hides a larger one, the second meta-narrative, that he calls the geopolitical narrative, in which the global disposition of centre-periphery contains an anthropological meaning and racist logic that classifies the whole of human life into humanity, non-humanity, or less-humanity (see also Fanon 1963; Gordon 2022). The third meta-narrative is the narrative of the myth of modernity (see Dussel 1993) that gives meaning to the capitalist and geopolitical narratives. This great meta-narrative, Bautista argues, has succeeded in penetrating the system of beliefs of the modern subject and its subjectivity structure. As mentioned earlier, the myth of modernity includes the idea of white European moral, political, scientific, and epistemic superiority that justifies the use of violence to reduce opposition or non-alignment. It also follows a linear conception of time, indeed a spatio-temporal hierarchy, according to which the West has the exceptional duty to forge the path to a developed future with its progressive ethos (Suárez-Krabbe 2016). Following Bautista, these three metanarratives maintain the hegemony of the social and economic system and create a world system through coloniality.

Systematically understanding and dismantling the negative moment makes possible the transition to the positive moment. This positive moment has two processes. The first is liberation from the world of modernity/coloniality, which means dismantling the structures of colonial domination that govern our subjectivities, belief systems, values, dreams and principles before we enter into the second process: the world of trans-modern rationality. This constitutes decoloniality *from*, whereas the second process constitutes decoloniality *for*. Both processes are epistemological and political. Bautista notes that the latter entails constructing a good life or, more precisely, is about moving towards *suma qamaña*, a Kichua concept that can be unpacked as follows:

> the *suma qamaña* is not restricted to material well-being, as expressed in the ownership of property or consumption at the heart of capitalist societies, but is a harmonious balance between material and spiritual components, which is only possible in the specific context of a community, which is social but also ecological. This social and ecological conception of community is linked to the Andean concept of the *ayllu*, where well-being encompasses not only persons, but also crops and cattle, and the rest of Nature. The classical Western dualism that separates society from Nature vanishes under this perspective, as one contains the other, and they are not separable.
>
> (Gudynas 2011, 444)

The *suma qamaña* is not alone in its uniqueness. There exists a vast plethora of life projects (Blaser 2019) that work against and beyond the death project of modernity (Suárez-Krabbe 2016), grounded in different understandings of personhood and community, but also nature, spirituality, and the cosmos (see also Escobar 2016). According to Bautista, the most difficult part of decolonisation is to expel modernity/coloniality within our subjectivity and rebuild humanity (see also Suárez-Krabbe 2022). Drawing inspiration from Paulo Freire, he urges us to use the following question pedagogically and methodologically: "How do we live liberation?" In this vein, we do not provide a manual or instructions on decolonisation, nor say that decoloniality and decolonisation must take the same paths. Many anticolonialists, anti-imperialists, and decolonialists have acted on these tasks in different ways throughout various periods of modern colonial history. Colonialism and coloniality are also differently articulated depending on the local histories of the places they are found. However, some ideas or paths can be named, as in the case of the path suggested by Bautista or those of many others that have struggled against this civilisation of death; that is the modern world era, its colonial designs, and the coloniality in which we live.

Notes

1 Drawing from Aronsson (2010, 558-9), we use the term Nordic (Norden) "because it unambiguously comprises Iceland, Denmark (with Greenland and the Faroe Islands), Norway, Sweden and Finland (with Åland)." Scandinavia is more ambiguous as it can refer to the geographical Scandinavian Peninsula (Sweden and Norway) with Denmark. Norden, however, can also become a complicated designation inasmuch, as Aronsson states, the Baltic countries, Russia, Northern Germany, and the diasporas in North America and elsewhere could also be counted as significant parts of the Nordic.
2 For a discussion of methods and methodologies in connection to Danish anthropology, see Suárez-Krabbe 2016, 144-154. See also Burman 2018.
3 Already at the end of the 1990s, Amanda Peralta, an Argentinian refugee working at the University of Gothenburg, was one of the scholars introducing postcolonial studies in Sweden, and specifically the decolonial perspectives of Walter Mignolo.
4 This concept, elaborated by anthropologist Ulla Vuorela and based on Gayatrik Spivak's conceptualisation of "complicity" (1999), was key in the volume edited by Suvi Keskinen et al. (2009) *Complying with colonialism.* Vuorela's article focuses on the role of Finland in the period of the European colonial expansion, where Finland was not part of the colonial project but neither an innocent victim of imperial pressures to adapt and reproduce colonial politics (Vuorela 2009, 19). She defines complicity as

> "participation in the hegemonic discourses, involvement in the promotion of universal thinking and practices of domination. Complicity is an important notion for those of us who are not quite situated in the centre; always wanting to get closer; our responses to the invitation give us a complicit position that we rarely even recognise. It also resembles a "tacit" acceptance of hegemonic discourses, since if we want to be accepted by the centres it is only possible, or so we think, on their terms".
>
> (Vuorela 2009: 20)

References

Aatar, Fátima. 2021. Prologo. Lo Peor Está por Venir. In *Intégrate Tú! Hablan los Indígenas de la República Francesa*, edited by H. Bouteldja and S. Khiari. Manresa: Ediciones Bellaterra.

Abdelhady, D., N. Gren, and M. Joormann, eds. 2020. *Refugees and the Violence of Welfare Bureaucracies in Northern Europe*. Manchester, England: Manchester University Press.

Abdel-Malek, Anouar. 1963. Orientalism in Crisis. *Diogenes* 11, no. 44: 103–140. https://doi.org/10.1177/039219216301104407

Achcar, Gilbert. 2008. Orientalism in Reverse. *Radical Philosophy* 151 (September/October 2008).

Adlbi Sibai, Sirin. 2016. *La Cárcel del feminismo – Hacia un Pensamiento Islámico Decolonial*. Madrid: AKAL/Inter Pares.

Ahmed, Sara. 2011. *Vithetens Hegemoni*. Hägersten: Tankekraft Förlag.

Ålund, Alexandra, and Carl-Ulrik Schierup. 1991. *Paradoxes of Multiculturalism: Essays on Swedish Society*. Avebury: Academic Publishing Group.

Ålund, Alexandra, Carl-Ulrik Schierup, and Anders Neergaard. 2017. *Reimagineering the Nation: Essays on Twenty-First-Century Sweden*. Sweden: Peter Lang Publishing Group.

Amadae, S. M., and C. Craig. 2022. Joining NATO in the Nuclear Age. *Helsinki Times*. https://www.helsinkitimes.fi/columns/columns/viewpoint/21618-joining-nato-in-the-nuclear-age.html

Andersen, Astrid, Kirsten Hvenegård-Lassen, and Ina Knobblock. 2015. 'Feminism in Postcolonial Nordic Spaces.' *Nora – Nordic Journal of Feminist and Gender Research* 23 (4): 239–45. https://doi.org/10.1080/08038740.2015.1104596

Andersen, Ole Stig, and René Mark Nielsen. 1987. *Noget fremmed: en bog om integration*. København: Forlaget Dünya.

Anderson, Perry. 1998. *The Origins of Postmodernity*. London: Verso.

Andersson, Roger and Irene Molina. 2018. Racialization and Migration in Urban Segregation Processes Key Issues for critical geographers. In *Voices from the North* (pp. 261–282). New York: Routledge.

Andreassen, Rikke and Myong, Lene. 2017. Race, Gender, and Researcher Positionality Analysed through Memory Work. *Nordic Journal of Migration Research* 7(2): 97–104.

Arce, José, and Julia Suárez-Krabbe. 2018. Racism, Global Apartheid and Disobedient Mobilities: The Politics of Detention and Deportation in Europe and Denmark. *KULT. Postkolonial Temaserie*, no. 15: 107–127. http://postkolonial.dk/wp-content/uploads/2017/09/11_Julia-og-Jose_We-are-here-because-you-were-there_final.pdf

Aronsson, Peter. 2010. Uses of the Past – Nordic Historical Cultures in a Comparative Perspective. *Culture Unbound* 2: 553–563. http://www.cultureunbound.ep.liu.se

Baldwin, James. 2017. *I am Not Your Negro*. New York: Vintage International.

Bannerji, Himani. 1995. *Thinking Through – Essays on Feminism, Marxism and Anti-Racism*. Toronto: Women's Press.

Bautista, Rafael. 2022. Retos Descoloniales en Educación, Investigación e Innovación. Conversatorio en Educación, Investigación e Innovación para la Liberación y el Buen Vivir. ComunasVE. Streamed on YouTube. https://www.youtube.com/watch?v=yCM_S3iRSFg

Beliso-De Jesús, M. A., and Jemima Pierre. 2020. Anthropology of White Supremacy. *American Anthropologist* 122, no. 1: 65–75.

Blaser, Mario. 2019. Life Projects. In *Pluriverse: A Post-Development Dictionary*, edited by Ashish Kothari, 234–236. New Delhi: Tulika Books.

Blaut, J.M. 1993. *The Colonizer's Model of the World – Geographical diffusionism and Eurocentric History*. New York: The Guilford Press.

Bouteldja, Houria, and Sadri Khiari. 2021. *¡Intégrate tú! Hablan los Indígenas de la República Francesa*. Manresa: Ediciones Bellaterra.

Burman, Anders. 2018. Are Anthropologists Monsters? An Andean Dystopian Critique of Extractivist Ethnography and Anglophone-Centric Anthropology. *HAU: Journal of Ethnographic Theory* 8, no. 1–2: 48–64. https://doi.org/10.1086/698413

Christensen, Julia, and Jens Heinrich. 2014. In Conversation: Shifting Narratives of Colonialism through Reconciliation in Greenland and Canada. *Kult*: 38–48. http://postkolonial.dk/files/KULT%2014/4%20Christensen%20and%20Heinrich%20final.pdf

Collins, Patricia Hill, and Sirma Bilge. 2016. *Intersectionality*. Cambridge: Polity.

Cox, Oliver C. 1959. *Caste, Class, & Race – A Study in Social Dynamics*. New York: Monthly Review Press.

Crenshaw, Kimberle. 1991. Mapping the Margins: Intersectionality, Identity Politics, and Violence Against Women of Color. *Stanford Law Review* 43: 1241–1299.

D'Amico-Samuels, Deborah. 1997. Undoing Fieldwork: Personal, Political, Theoretical and Methodological Implications. In *Decolonizing Anthropology:*

Moving Further Toward an Anthropology for Liberation, edited by Faye Harrison, 68–87. Arlington, VA: Association of Black Anthropologists – American Anthropological Association.

Dahl, Ulrika. 2021. Nordic Academic Feminism and Whiteness as Epistemic Habit. In *Feminisms in the Nordic Region. Gender and Politics*, edited by S. Keskinen, P. Stoltz, and D. Mulinari. Palgrave Macmillan, Cham. https://doi.org/10.1007/978-3-030-53464-6_6

Davis, Angela. 1981. *Women, Race & Class*. 1st ed. New York: Random House.

De los Reyes, Paulina, Irene Molina, and Diana Mulinari. 2002. *Maktens (o)lika förklädnader. Kön, klass och etnicitet i det postkoloniala Sverige*. Stockholm: Atlas.

De los Reyes, Paulina, and Diana Mulinari. 2020. Hegemonic Feminism Revisited: On the Promises of Intersectionality in Times of the Precarisation of Life. *NORA-Nordic Journal of Feminist and Gender Research* 28, no. 3: 183–196.

Dirlik, Arif. 1994. The Postcolonial Aura: Third World Criticism in the Age of Global Capitalism. *Critical Inquiry* 20, no. 2: 328–356.

Dussel, Enrique. 1993. Eurocentrism and Modernity (Introduction to the Frankfurt Lectures). *Boundary* 2, no. 3: 65–76.

Dussel, Enrique. 1994. *1492 – El Encubrimiento del Otro: Hacia el Origen del "Mito de la Modernidad"*. Mexico: Ediciones Abya-Yala.

Echeverrria, Bolívar. 2010. *Modernidad y Blanquitud*. Tlalpan: Biblioteca Era.

Escobar, Arturo. 2016. Thinking-feeling with the Earth: Territorial Struggles and the Ontological Dimension of the Epistemologies of the South. *Revista de Antropología Iberoamericana* 11, no. 1: 11–32. 10.11156/aibr.110102e

Espinosa, Miñoso Yuderkys. 2009. Etnocentrismo y Colonialidad en los Feminismos Latinoamericanos: Complicidades y Consolidación de las Hegemonías Feministas en el Espacio Transnacional. *Revista Venezolana de Estudios de la Mujer* 14, no. 33 37–54.

Fals Borda, Orlando. 1979. *El Problema de Cómo Investigar la Realidad para Transformarla por la Praxis*. Bogotá: Ediciones Tercer Mundo.

Fals Borda, Orlando. 1981. *Ciencia propia y colonialismo intelectual*. Bogotá: Carlos Valencia Editores.

Fanon, Frantz. 1963. *The Wretched of the Earth*. New York: Grove.

Fanon, Frantz. 1967. *Black Skins White Masks*. ed. Charles Lam Markman. London: Grove Press.

Finck Carrales, Juan Carlos, and Julia Suárez-Krabbe, eds. 2022. *Transdisciplinary Thinking from the Global South: Whose Problems, Whose Solutions?* London: Routledge. Routledge Research on Decoloniality and New Postcolonialisms. https://doi.org/10.4324/9781003172413

Goldberg, David T. 2002. *The Racial State*. Oxford: Blackwell Publishers Ltd.

Gordon, Lewis. 2006a. African-American Philosophy, Race, and the Geography of Reason. In *Not Only the Master's Tools: African-American Studies in Theory and Practice*, edited by Lewis Gordon and Jane Anna Gordon, 3–50. Boulder, CO and London: Paradigm.

Gordon, Lewis. 2006b. *Disciplinary Decadence: Living Thought in Trying Times*. Boulder, CO and London: Paradigm.

Gordon, Lewis. 2011. Shifting the Geography of Reason in an Age of Disciplinary Decadence. *Transmodernity* 1, no. 2: 95–103.

Gordon, Lewis. 2021. *Freedom, Justice, and Decolonization*. New York: Routledge.

Gordon, Lewis. 2022. *Fear of Black Consciousness*. New York: Farrar, Straus and Giroux.

Graugaard, Naja Dyrendom. 2014. Uanga ('I'): Journey of Raven and the Revival of the Spirit of Whale. *Kult*: 6–22. http://postkolonial.dk/files/KULT%2014/2%20Uanga%20(I)%20-%20Naja%20Dyrendom%20Graugaard.pdf
Greve, Bent. 2016. Migrants and Health in the Nordic Welfare States. *Public Health Reviews* 37, no. 9. https://doi.org/10.1186/s40985-016-0023-6
Groglopo, Adrián. 2012. *Appropriation by Coloniality: TNCs, Land, Hegemony and Resistance. The Case of Botnia/UPM in Uruguay*. Doctoral dissertation, Umeå University.
Groglopo, Adrián. 2015. Antirasistiska Akademins Uppkomst – En Historia om Politisk Kamp i Vetenskapens idéfält. In. *Vardagens Antirasism: om Rörelsens villkor och Framväxt i Sverige*, edited by A. Groglopo, M. Allelin, D. Mulinari, and C. Diaz. Lund: Antirasistiska Akademin.
Groglopo, Adrián and Beth M. Ahlberg. 2006. Hälsa, Vård och Strukturell Diskriminering. *Hälsa, Vård och Strukturell Diskriminering [Health, Care and Structural Discrimination]. SOU 2006: 78*. Stockholm: Fritzes.
Grosfoguel, Ramón. 2018. ¿Negros Marxistas o Marxismos Negros? Una Mirada Descolonial. *Tabula Rasa* 28: 11–22. https://doi.org/10.25058/20112742.n28.1.
Grosfoguel, Ramón. 2011. Decolonizing Post-Colonial Studies and Paradigms of Political-Economy: Transmodernity, Decolonial Thinking, and Global Coloniality. *TRANSMODERNITY: Journal of Peripheral Cultural Production of the Luso-Hispanic World*.1: 1.
Grosfoguel, Ramón. 2013. The Structure of Knowledge in Westernised Universities: Epistemic Racism/Sexism and the Four Genocides/Epistemicides. *Human Architecture: Journal of the Sociology of Self-knowledge* 1, no. 1: 73–90.
Gudynas, E. 2011. Buen Vivir: Today's Tomorrow. *Development* 54: 441–447. https://doi.org/10.1057/dev.2011.86
Gullestad, Marianne. 2004. 'Blind Slaves of Our Prejudices: Debating "Culture" and "Race" in Norway'. *Ethnos* 69 (2): 177–203. https://doi.org/10.1080/0014184042000212858
Habel, Ylva. 2012. Challenging Swedish Exceptionalism? Teaching While Black. *Education in the Black Diaspora* 2012. 110–133.
Hall, Stuart. 1980. Race, Articulation and Societies Structured in Dominance. In *Sociological Theories: Race and Colonialism*, edited by Unesco, 305–345. Paris: Unesco.
Hervik, Peter. 2015. 'Race, "Race", Racialisering, Racisme og Nyracisme'. *Dansk Sociologi* 26 (1): 29–50.
Hervik, Peter, ed. 2019. *Racialization, Racism, and Anti-racism in the Nordic Countries*. New York: Palgrave Macmillan.
Höglund, Johan, and Linda Andersson Burnett. 2019. Introduction: Nordic Colonialisms and Scandinavian Studies. *Scandinavian Studies* 91, nos. 1–2: 1–12. Web.
Hübinette, Tobias and Catrin Lundström. 2014. Three Phases of Hegemonic Whiteness: Understanding Racial Temporalities in Sweden. *Social Identities* 20, no. 6: 423–437.
Icaza, Rosalba, and Rolando Vázquez. 2018. Diversity or Decolonisation? Researching Diversity at the University of Amsterdam. In *Decolonising the University*, edited by Gurminder K. Bhambra, Dalia Gebrial, and Kerem Nisancioglu. London: Pluto Press.
James, C. L. R. 1989 [1963, 1938]. *The Black Jacobins: Toussaint L'Ouverture and the San Domingo Revolution*. 2nd ed. New York: Vintage Books.

Jensen, Lars. 2018. *Postcolonial Denmark: Nation Narration in a Crisis Ridden Europe*. Abingdon, Oxon; New York, NY: Routledge. Print.

Kamali, Masoud. 2021. *Neoliberal Securitisation and Symbolic Violence: Silencing Political, Academic and Societal Resistance*. Cham: Springer Nature.

Karatani, Kojin. 2018. Neoliberalism as a Historical Stage. *Global Discourse* 8, no. 2: 191–207.

Keskinen, Suvi. 2016. From Welfare Nationalism to Welfare Chauvinism: Economic Rhetoric, the Welfare State and Changing Asylum Policies in Finland. *Critical Social Policy* 36 (3): 352–70. https://doi.org/10.1177/0261018315624170

Keskinen, Suvi, Unnur Dís Skaptadóttir, and Mari Toivanen, eds. 2019. *Undoing Homogeneity in the Nordic Region: Migration, Difference and the Politics of Solidarity*. Studies in Migration and Diaspora. Abingdon, Oxon; New York, NY: Routledge.

Keskinen, Suvi, Pauline Stoltz, and Diana Mulinari, eds. 2021. *Feminisms in the Nordic Region. Neoliberalism, Nationalism and Decolonial Critique*. London: Gender and Politics. Palgrave Macmillan. https://doi.org/10.1007/978-3-030-53464-6_6

Keskinen, Suvi, Salla Tuori, Sari Irni, and Diana Mulinari. 2009. *Complying with Colonialism: Gender, Race, and Ethnicity in the Nordic Region*. Farnham, UK: Ashgate.

Khatibi, Abdelhebir [1981] Décolonisation de la sociologie. In *Chemins de traverse : essais de sociologie*, textes réunis et revus par NejjarS., edited by A. Khatibi. 113–125. Rabat, Éditions Okad.

Khawaja, Iram. 2015. "Det muslimske sofa-hjørne": Muslimskhed, Racialisering OG Integration. *Pædagogisk Psykologisk Tidsskrift* 52 (2): 29–38.

Kohan, Nestor. 2022. Crítica y Polémica en el Vacío Posmoderno. In *La Isla posible. Ensayos sobre Ideología y Revolución*, edited by E. Ubieta. Buenos Aires: Ediciones Acercándonos.

Kothari, Ashish; Salleh, Ariel; Escobar, Arturo; Demaria, Federico and Acosta, Alberto. 2019. Introduction. Finding Pluriversal Paths, In *Pluriverse. A Post-Development Dictionary*, XXI–XXXV. New Delhi: Tulika Books.

Lauesen, Torkil. 2021. *Riding the Wave – Sweden's Integration into the Imperialist World System*. Montreal: Kersplebedeb.

Loftsdóttir, Kristin, and Lars Jensen, eds. 2012. *Whiteness and Postcolonialism in the Nordic Region*. Abingdon, Oxon: Taylor and Francis.

Lugones, María. 2003. *Pilgrimages/Peregrinajes: Theorizing Coalition Against Multiple Oppressions*. Lanham: Rowman and Littlefield.

Lugones, Maria. 2010. The Coloniality of Gender. In *The Transgender Studies Reader Remix*, edited by S. Stryker and S. Whittle. 144–156. New York, NY: Routledge, 2022.

Maldonado-Torres, Nelson. 2020. What is Decolonial Critique? *Graduate Faculty Philosophy Journal* 41, no. 1: 157–183.

Martínez, Elizabeth and Davis, Angela Y. 1994. Coalition Building Among People of Color. *Inscriptions*. **7**: 42–53.

McEachrane, Michael, ed. 2014. *Afro-Nordic Landscapes: Equality and Race in Northern Europe*. Routledge Studies on African and Black Diaspora 5. New York: Routledge, Taylor & Francis Group.Mearsheimer, John J. 2022. Playing With Fire in Ukraine – The Underappreciated Risks of Catastrophic Escalation. *Foreign Affairs*. https://www.foreignaffairs.com/ukraine/playing-fire-ukraine

Meiksins Wood, Ellen. 1986. *The Retreat from Class – A New 'True' Socialism*. London: Verso.

Mignolo, Walter D. 2000. *Local Histories/Global Designs. Coloniality, Subaltern Knowledges, and Border Thinking*. New Jersey: Princeton University Press.

Mignolo, Walter D. 2011. *The Darker Side of Western Modernity: Global Futures, Decolonial Options*. Durham & London: Duke University Press.

Mignolo, Walter D., and Catherine E. Walsh. 2018. *On Decoloniality*. Durham & London: Duke University Press.

Monahan, Michael. 2011. *The Creolizing Subject. Race, Reason and the Politics of Purity*. New York: Fordham University Press.

Montañez Pico, Daniel. 2020. *Marxismo Negro – Pensamiento Descolonizador del Caribe Anglófono*. Mexico: Akal/Inter Pares.

Moorhead, Shelley. 2017. *Whiteness, Knowledge Production and Politics*. Keynote at Roskilde University. March 20, 2017. http://postkolonial.dk/kult-15-racism-in-denmark/

Mulinari, Diana and Anders Neergaard. 2022. The Swedish Racial Welfare Regime in Transition. In *Racism in and for the Welfare State. Marx, Engels, and Marxisms*, edited by F. Perocco. Cham: Palgrave Macmillan. https://doi.org/10.1007/978-3-031-06071-7_4

Mulinari, Leandro Schclarek, and Suvi Keskinen. 2020. 'Racial Profiling in the Racial Welfare State: Examining the Order of Policing in the Nordic Region'. *Theoretical Criminology* 1 (19): 1–19. https://doi.org/10.1177/1362480620914914

Myong, Lene, and Mons Bissenbakker. 2016. 'Love without Borders?' *Cultural Studies* 30 (1): 129–46. https://doi.org/10.1080/09502386.2014.974643

Naum, Magdalena, and Jonas M. Nordin, eds. 2013. *Scandinavian Colonialism and the Rise of Modernity: Small Time Agents in a Global Arena*. New York: Springer Science & Business Media.

Nkrumah, Kwame. 1970. *Class Struggle in Africa*. New York: International Publishers.

Oyěwùmí, Oyèrónké. 1997. *The Invention of Women: Making an African Sense of Western Gender Discourses*. Minneapolis: Univ. of Minnesota Press.

Padovan-Özdemir, Marta, and Øland, Trine. 2022. *Racism in Danish Welfare Work with Refugees: Troubled by Difference, Docility and Dignity*. Abingdon, Oxon; New York, NY: Routledge. https://doi.org/10.4324/9781003097327

Quijano, Aníbal. 2000. Coloniality of Power, Ethnocentrism, and Latin America. *Nepantla, Views from the South* 1, no. 3: 533–580.

Reddock, Rhoda E. 1994. *Women, Labour and Politics in Trinidad and Tobago: A History*. Kingston: Ian Randle.

Rivera, Cusicanqui Silvia. 2012. Ch'ixinakax Utxiwa: A Reflection on the Practices and Discourses of Decolonization. *The South Atlantic Quarterly* 111, no. 1: 95–109.

Robinson, Cedric J. 1983. *Black Marxism. The Making of the Black Radical Tradition*. Carolina: University of North Carolina Press.

Rodney, Walter. 1972. *How Europe Underdeveloped Africa*. London: Bogle-L'Ouverture.

Romero-Losacco, José. 2018. *La invención de la Exclusión – Individuo, Desarrollo e Inclusión*. Caracas: Fundación Editorial El Perro y la Rana.

Sandoval, Chela. 2000. *Methodology of the Oppressed*. Minneapolis: University of Minnesota Press.

Santos, Boaventura de Sousa. 2014. *Epistemologies of the South: Justice Against Epistemicide*. London: Routledge.

Sawyer, Lena and Ylva Habel. 2014. Refracting African and Black diaspora through the Nordic region, *African and Black Diaspora: An International Journal* 7, no. 1: 1–6. https://doi.org/10.1080/17528631.2013.861235

Schierup, Carl-Ulrik, Peo Hansen and Stephen Castles. 2006. *Migration, Citizenship, and the European Welfare State: A European Dilemma*. Oxford: Oxford University Press.

Segato, Rita L. 2016. Patriarchy from Margin to Center: Discipline, Territoriality, and Cruelty in the Apocalyptic Phase of Capital. *South Atlantic Quarterly* 115, no. 3: 615–24.

Small, Stephen. 2018. *20 Questions and Answers on Black Europe*. The Hague: Amrit Publishers.

Suárez-Krabbe, Julia. 2016. *Race, Rights and Rebels: Alternatives to Human Rights and Development from the Global South*. New York & London: Rowman & Littlefield International. Global Critical Caribbean Thought.

Suárez-Krabbe, Julia. 2022. Over Our Dead Bodies: The Death Project, Egoism and the Existential Dimensions of Decolonization. In *Transdisciplinary Thinking from the Global South: Whose Problems, Whose Solutions?*, edited by J. C. F. Carrales and J. Suárez-Krabbe, 130–147. New York: Routledge Research on Decoloniality and New Postcolonialisms. https://doi.org/10.4324/9781003172413-7

Suárez-Krabbe, Julia, and Annika. Lindberg. 2019. Enforcing Apartheid? The Politics of Intolerability in the Danish Migration and Integration Regimes. *Migration and Society. Advances in Research* 2, no. 1: 90–97. https://doi.org/10.3167/arms.2019.020109

Suárez-Krabbe, Julia. 2012. 'Epistemic Coyotismo' and Transnational Collaboration: Decolonizing the Danish University. *Human Architecture: Journal of the Sociology of Self-Knowledge* 10, no. 1, 5. http://scholarworks.umb.edu/humanarchitecture/vol10/iss1/5

Suárez-Krabbe, Julia. 2014. Pluriversalizing Europe: Challenging Belonging, Revisiting History, Disrupting Homogeneity. *Postcolonial Studies* 17, no. 2: 155–172. https://doi.org/10.1080/13688790.2014.966413

Tafari-Ama, Imani. 2020. An African Caribbean Perspective on Flensburg's Colonial Heritage. *Kult*: 7–30. http://postkolonial.dk/wp-content/uploads/2020/10/1-Imani-article-final.pdf

Tesfahuney, Mekonnen. 2005. Uni-versalism. In I. P. de los Reyes & M. Kamali (red.). *Bortom vi och dom: Teoretiska reflektioner om makt, integration och strukturell diskriminering*, 203–232. Stockholm: Fritzes. 41.

Tesfahuney, Mekonnen and Lena Grip. 2007. Integrationism: Viljan till detsamma. *Nordisk Samhällsgeografisk Tidskrift* 43: 61–91.

Tuhiwai-Smith, Linda. 1999. *Decolonizing Methodologies: Research and Indigenous Peoples*. London: Zed Books.

Vasco, Luis Guillermo. 2002. *Entre Selva y Páramo. Viviendo y Pensando la Lucha India*. Bogotá: Instituto Colombiano de Antropología e Historia.

Vasco, Luis Guillermo. 2007. Así Es Mi Método en Etnografía. *Tabula Rasa* 6:19–52.

Velásquez, Atehortúa Juan. 2020. A Decolonial Pedagogy for Teaching Intersectionality. *NJCIE: Nordic Journal of Comparative and International Education* 4, no. 1: 156–171.

Vergès, Françoise. 2021. *A Decolonial Feminism*. Northamptom: Pluto Press.

Vitus, Kathrine, and Rikke Andreassen. eds. 2015. *Affectivity and Race: Studies from Nordic Contexts. Farnham, Surrey, England; Burlington*, London: Ashgate Publishing.

Vuorela, Ulla. 2009. Colonial Complicity: The 'Postcolonial' in a Nordic Context. In *Complying with Colonialism: Gender, Race, and Ethnicity in the Nordic Region*, edited by Suvi Keskinen, Salla Tuori, Sari Irni, and Diana Mulinari, 19–33. Farnham, UK: Ashgate.

Wa Thiong'O, Ngugi. 1986. *Decolonising the Mind: The Politics of Language in African Literature*. Nairobi: East African Educational Publishers Ltd.

Wallace, Matthew, Michael J. Thomas, José Manuel Aburto, Anna Vera Jørring Pallesen, Laust Hvas Mortensen, Astri Syse et al. 2021. *The Impact of the Mortality of International Migrants on Estimates and Comparisons of National Life Expectancy: A Comparative Study of Four Nordic Nations*. Stockholm: Stockholm Research Reports in Demography. https://doi.org/10.17045/sthlmuni.14763243.v1

Wallerstein, Immanuel. 1996. *Open the Social Sciences: Report of the Gulbenkian Commission on the Restructuring of the Social Sciences*. Palo Alto, CA: Stanford University Press.

Walsh, Catherine. 2018. The Decolonial For: Resurgences, Shifts, and Movements. In *On Decoloniality*, edited by Walter Mignolo and Catherine Walsh, 15–32. Durham & London: Duke University Press.

Williams, Eric E. 1944. *Capitalism and Slavery*. London: A. Deutsch.

Zakaria, Rafia. 2021. *Against White Feminism. – Notes on Disruption*. London: W.W. Norton & Company Ltd.

1 Surviving like Scheherazade. Veiled women and liberalism

The trap of the progressive left[1]

Houria Bouteldja

May Allah protect us from the word 'I'

Yesterday the forced unveiling, today the voluntary unveiling

A standing ovation greets the public unveiling of sociologist Zahra Ali, an emblematic figure of Islamic feminism in Europe, on the day that she posts a picture of herself with uncovered hair on her Facebook page. The roaring applause does not come from well-known Islamophobes but from intersectional feminists, anti-racists, progressives, and activists who are sincerely committed to the struggle against Islamophobia; these are my friends and comrades too. So, what elicits their admiration? It is Zahra Ali's capacity (according to them) to have made a genuine choice, to have exercised her freedom as a woman, to have had the courage to challenge both Islamophobes and her own community, that made her a feminist *par excellence*. She reinstated the notion of free will and individualism through her action. In other words, her unveiling symbolizes a dazzling achievement of the feminist ideal: the capacity to exercise full sovereignty over one's body and mind.

Among the Muslim community, and for those who saw her as a role model, it is a sentiment of consternation that prevails. Indeed, by fighting in the public arena, by proudly wearing the veil and by successfully demonstrating that you could wear it and follow your faith while being a free woman, Zahra Ali gave countless women who wear the veil the dignity and respect that public debate had deprived them of for so long. Rather than an emancipatory act, disappointed Muslims (men and women) saw this as a form of treason.

I have to admit that I was one of them. But mostly, I felt perplexed. Not so much because she removed her veil—that is absolutely her right to do so; I am no conscience keeper and I do not wear the veil myself. Instead, it is the public nature of her action that is relevant here because it politicised it. She proved many white progressives, who had originally opposed the French law of 15 March 2004,[2] right when they argued to oppose this Islamophobic bill while also highlighting the oppressive nature of the veil. To them, supporting Muslim women and they start being exposed to real emancipation; they will end up removing the veil themselves. So, when I saw Zahra Ali publicised and

DOI: 10.4324/9781003293323-2

proudly claim what she had done, I told myself: "the progressives got us." Since then, other Islamic (or Muslim) feminists have reached the same conclusion, such as activists from the French group Lallab, for instance: Leila Alaouf,[3] the sociologist Hanane Karimi, or the Moroccan intellectual Asma Lamrabet.[4]

Some (idealist) readers may be surprised by my lack of empathy towards a "sister" and in light of an act that some would see as courageous (and in some ways it was, when you see the hostility it attracted from a majority of Muslims). They might even argue that her freedom is sacred and that she does not have to compromise because of her community, which must surely be sometimes (or often) oppressive.

This is likely the position that most progressives will take irrespective of the context, glossing over the colonial dimension of power relations on a global scale, thereby also putting the racist French system on the same level as the specific and particular behaviour of the Muslim community, which I would like to emphasise is a social, political, and institutional minority in France, meaning it remains cut off from all levers of power. But in a context of triumphant liberalism in all its forms, it is hard for a materialist approach to power relations to be heard, and it has failed to grapple with the emergence of Islamic feminism whose most ready incarnation seems to be liberal progressivism.

It is worth pausing to consider the ideological underpinnings of this kind of feminism where the (voluntary) unveiling of some of its representatives in Western countries strangely extends (forced) unveiling under colonial rule. Of course, Islamic feminists would contest this interpretation, especially since they have all endorsed anti-racist and anticolonial critiques. Yet both positions mobilise the same logic of "women liberation", albeit one for racist purposes and the other for progressive ones. So, for me, we should not see this act of unveiling as a random accident or contradiction, but as a logical and natural progression, almost the unavoidable endpoint of this feminism, even. In order to provide further evidence for this point, it is important to foreground the Western-centric paradigm that underlines Islamic feminism, itself embedded in the reification of Muslim history.

Islam, a feminist religion?

Islam is a feminist religion and the "choice" of wearing the veil is an Islamic version of the famous slogan, "my body, my choice" (or "my body belongs to me" in the French rendition of the English slogan). At least, this is what some Islamic feminists argue in the anthology *Islamic Feminisms*, edited by Zahra Ali (2020). The anthology brings together the most prestigious and well-known figures of this movement, including Asma Barlas, Margot Badran, Asma Lamrabet, Malika Hamidi, and Ziba Mir-Hosseini; it also draws the theoretical and conceptual contours of Islamic feminism very precisely and, as such, the book can legitimately be used as the basis for critique of the movement.

There is no doubt in my mind that Islam is a more egalitarian religion[5] than other monotheistic religions, for reasons that also figure prominently in Zahra Ali's anthology (2020):

- In the sacred Quran, Eve is not born from Adam's rib and thus does not come second in Creation. She does not bear any responsibility for the original sin either. However, she shares with man the same duty of personal piety upon which all human beings are eventually judged.
- Man's superiority over woman stands in complete contradiction with divine revelation to the extent that it goes against the notion of "Tawhid", which stipulates that Allah's sovereignty is indivisible and cannot be shared with anyone. Therefore, no man can claim this attribute of superiority as it would equate with claiming a divine attribute.
- Unlike the Christian tradition, Allah is never represented. He is not born from someone and does not give birth to anyone. As a result, he cannot be represented in a masculine form which would strengthen masculine domination over women. Non-gendered, Allah does not promote any gender.
- Lastly, there is consensus among Islamic feminists that sexist interpretations of the Quran and the Sunna are the product of men through the ages. I tend to agree with this interpretation considering that the patriarchal character of the regions where Islam first arose and took hold is well known.

Following from this, the masculine interpretation of the Quran and the Sunna is first and foremost a result of power relations in societies that were already largely structured by patriarchal power. As such, in the context of Muslim societies today (with traditional patriarchy + colonial heterosexism structuring the terrain of power and politics), a feminist reading of the Quran is a political move which takes aim at patriarchal forces with the only legitimate currency available: the Quran. I think it is easy to conclude, without wanting to offend anyone, that this is an instrumental way of going about things (the Quran is a means to an end in this process)—which is exactly what men have done to establish their domination over women. In this regard, Ziba Mir-Hosseini (2020, 130) acknowledges this herself when she writes: "By using the language of political Islam, Islamic feminists have managed to critique sexist prejudices in Islamic law using means that were previously unimaginable."

I believe this is a legitimate approach, but this is also where I stop agreeing with Islamic feminists. While it is one thing to use sacred scriptures strategically to pursue feminist goals, it is another to assert that Islam is a "feminist" religion since the time of Revelation. It is not enough for the Quran to posit the equal dignity of men and women to turn Islam into a "feminist" religion. This is completely anachronistic. It is absurd to say Islam is "feminist"—as much as it would be absurd to call it "Marxist," "communist," "libertarian," "fascist," or "reactionary"—simply because these movements are the product of Western modernity. Claiming that Islam is feminist is occulting the material conditions that gave rise to feminism in Europe. It is turning feminism into a timeless, ahistorical phenomenon. It is not enough to remind us that Aisha[6] was a warlord, or that women substantially contributed to scientific knowledge and discovery at that time, to make Islam a feminist religion.

In the West, feminism is a major political phenomenon with its theorists, its movements, and competing trends, claims and programmes—all of which aim to abolish patriarchy within the context of democratic societies that value equality and universalism, and with capitalist expansion providing the material means to turn these ideals (of principled equality and the likes) into facts. Equality itself, as an essential principle of political philosophy, comes from the French Revolution and was later enshrined in the French constitution in the 20th century. Hence, feminism in this sense is a highly specific phenomenon, anchored in Western societies during modernity. The history of Islam has its own singular characteristics, and it does not make sense to want to match it to that of the colonial West, unless one persists in raising European political phenomena to such a degree of universality and timelessness that not claiming it would be tantamount to renouncing one's own dignity?

Islamic Feminisms (2020) loses its credibility because of this sort of anachronism. The authors use overbearing assertions to push their argument at all costs, as if anticipating the recalcitrance of their audience (both white and Muslim), and this raises even more questions and suspicions. As such, I want to examine more closely the critiques that have been levelled at Islamic feminism.

Islamic feminism trapped in Western modernity

Sirin Adlbi Sibai's (2016) remarkable book *La Cárcel del feminismo – Hacia un pensamiento islámico decolonial* [*The jail of feminism – Towards a decolonial Islamic thought*] is a very valuable resource here. In it, she describes how, since the rise of Western feminism, Islam has been trapped in the binary opposition tradition/modernity as the main colonial conceptual apparatus of what she calls "the epistemological/existential prison." Drawing on Moroccan philosopher Taha Abderrahman, she explains: "Colonised Arab-Muslim thinking emerged in the 18th century by integrating the binaries of tradition/modernity, identity/alterity, religious/secular as its central paradigm, culminating with the formation of independent Arab nation-states" (Adlbi 2016, 88). Because of its material superiority (the result of capital accumulation combined with military hegemony), the supposedly universal Western epistemology managed to wipe out the epistemology intrinsic to colonised and dominated societies. Historically, this epistemological loss of the colonised countries is why and how the Islamic tradition came to be dependent upon the Western epistemological framework, losing its singularity. From then, and as Ramón Grosfoguel points out, Islam borrowed from Christianity its binary vision of earthly/spiritual matters and faith/reason;

and from modernity, the dichotomy between public and private (Grosfoguel 2013, see also Adlbi 2016, 96). The Christian-centric processes of domestication of Islam in its secular form is still ongoing and continues to reproduce the conditions of Islam's own alienation. Indeed, Islam is not a "religion," according to the Andalusian intellectual Abdelmumin Aya, but a "Din," that is to say a tradition which refuses to separate the religious from the state (Adlbi 2016, 98).

Despite Margot Badran's argument that "Islamic feminism transcends and dismantles old binary divisions between religious/secular or East/West" (2020, 45), it does not escape this trap. On the contrary, it takes global secularisation as a sign of progress and the universalism of women's condition as a given. As Asma Lamrabet asserts, "the second evidence to put forward is the 'universality' of discriminations against women" (2020, 57).

Even though it is true that the condition of "the second sex" (Beauvoir 1953) has drastically deteriorated in the last two centuries (at least), and that there is a "world war" against women, it is useful to bear in mind that this is not a "natural" or "cultural" progression of events, but rather the result of globalisation as colonial expansion for capital accumulation. From a decolonial point of view that centres material power relations on an international scale, as well as their distinct and localised impacts, we cannot underestimate the ethnocentrism of white feminism and its universalist aspirations. Indeed, decolonial thought reveals that non-white women are social subjects who are oppressed through structural mechanisms of domination shaped by race, class, poverty, war, colonialism, or imperialism. Even though Islamic feminists include colonialism and its impact in their theoretical approach, they over-determine the role of "the patriarchal interpretation" of the Quran. In effect, they almost blame global colonial power (and its tentacular hegemony) equally—and as much as—localised patriarchal structures and powers (distinguishing the intrinsic from the changes that Western modernity brought upon these local structures is tenuous at best).

Asma Lamrabet adds:

> "We have to move away from *both* a hegemonic western political ideology and a rigorist or extremist vision of religion in the Muslim world, which are both positioned within a narrative of clash of civilisation and the rejection of the Other. This would finally open an alternative pathway for a realistic take on the complexity of women's condition in Islam" .
>
> (Lamrabet 2020, 59)

Asma Barlas goes even further:

> You can only blame history or the West to an extent when it comes to the oppression of women in Muslim countries; I think a lot of these issues come from a misinterpretation of the teachings of the Quran by Muslims.
>
> (Barlas 2020, 83)

Ziba Mir-Hosseini (2020, 128), offers a more nuanced and lucid point of view however: "The rise of modern nation-states in the Muslim world and their integration of Islam and of 'the woman question' in the nation-building process has made the situation a lot more complex". Yet, she goes on to point to "despotism" as the actual cause behind the condition of Muslim women, and hails "democracy" and a non-patriarchal reading of the Quran as the solution, weakening her overall argument. Obviously, it is not an entirely erroneous interpretation, but what about the role of neoliberalism and imperialism in the production of gendered relations? Mir-Hosseini's argument is all the more puzzling since she seems to rejoice that, three decades after the Iranian revolution, "society is more 'secular' now than it was before the revolution" (2020, 118).

In this perspective, Islamic feminism appears as one more expression of Islamic reformism that takes as its point of departure the idea that Islam is behind Western countries and under-developed, and that progress necessarily means secularisation. Despite genuine efforts to break away from the Western epistemological matrix, I find Islamic feminism lacking here. Its endorsement of Islamic reformism echoes the way Arab-Muslim nation states first utilised reformism, as an ideology to compete with big colonial powers and "stay in the race." In doing so, colonised people copied and implemented legal and political frameworks borrowed from white nationalism: unless they adopt a radically decolonial approach, feminisms and other emancipatory struggles are bound to follow the same path today and reproduce emancipatory approaches developed by white people. Emancipatory approaches developed *by* white people mean that they are also designed *for* white people, hence they require oppression. Like the Western freedom being dependent upon the colonised's unfreedom.

Furthermore, and at its core, Islamic feminism falters because of its idealism. What *really* drives the unenviable fate of women in Muslim societies is not so much this masculine appropriation of sacred texts (even though it is a factor), but the institutionalisation and incorporation of Islam into nation-building processes which borrowed so much from Western nationalism and its heterosexism. As Sirin Adlbi Sibai also notes: "Some Muslim regimes tried really hard to nationalize and institutionalize Islam, and obviously this did not contribute to reverse the trend of outrageous disadvantage for women in almost all areas of life, economically, socially, politically and culturally" (Adlbi Sibai 2016, 111).

Therefore, the new forms of Arab-Muslim patriarchy that emerged in the nation state building process were largely shaped by these dynamics, and I think it would be apt to describe these new models of masculinities as white and Christian-influenced masculinities. Indeed, these masculinities are the product of two concurrent pressures: Western heterosexism and its imposed binary thinking, on the one hand; and Muslim men's fears of losing power over women as a result of the "disloyal" competition coming from a super powerful West in an Islamic world imprisoned by ideals of modernity, on the other. Nadia Tazi (Mormin-Chauvac 2019) similarly concludes that the success of political Islam[7] in the Arab world is not so much the result of a religious push but of men doubling down on masculine/virile performance in the face of the transformations brought upon by modernity, dictatorial regimes, and Western hegemony. This "domineering over-the-top male" that we see on the street is a façade to dissimulate the "sub-male" as it actually exists in global power relationships. It is through this lens that we must understand the transformation and the strengthening of patriarchal structures in the Arab-Muslim world. It is also through this lens that we must fight the Family Code in Algeria, for example, because it is an extension of the heteronormativity and male domination embedded in the DNA of the patriarchal nationalism inherited from Western colonial structures. If it really must use the Quran to pursue feminist goals, I think Islamic feminism would be better off embracing a political materialist approach to Islam that sees its *contemporary expressions* (whether they be institutional or ideological) not as an essence, but as a historical production born out of the infrastructure of modernity.

Following from this, I have to agree with Nadia Yasin when she says, "the use of the concept of 'Islamic feminism' is really mostly a strategy" (quoted in Adlbi Sibai 2016, 264). Sirin Adlbi Sibai concurs: "[Islamic feminism] is the discursive and conceptual form used by Muslim women, consciously and unconsciously, when they seek a way to express themselves and be heard in the context of an epistemological monologue" (Adlbi Sibai 2016, 265). One can add that such an epistemological monologue is one between the West and its non-white mimes. And so, we need to ask ourselves: is this really a stable place to proceed from, or more of a deceiving quicksand?

The integration of Islam in the French (colonial) Republic and the individualisation of Muslims

What I turn to now is the notion of the individual as a concept in modern society, and its irresistible power and traction in enabling integration. Today, we hear that we are all individuals: there are no people, no nations, no classes, no groups; only free and selfish individuals, every one of them distinct and separate. In this perspective, the individual represents the biggest obstacle to collective mobilisation.

On the relationship between individuals and modern states, Sadri Khiari writes:

> "Individual freedom and political equality are the basic principles of capitalist democracies. Races are the negation of these principles. And yet they are inseparable from them. The bourgeois modernity that took hold at the turn of the 18th century and in the 19th century, is the fruit of two contradictory but complementary movements: individuals freeing themselves from the stronghold of statutory hierarchies—which was a necessary step for the formation of the modern state and the flourishing of Capital—and imperial expansion which was just as crucial."
>
> (Khiari 2009, 36)

Joan W Scott adds:

> These days, in the era of globalization, all aspects of life have become increasingly 'marketized' and the state's role is narrowed to a protector of market forces and individual self-determination. Society is conceived to be a mass of self-actualizing individuals, their fortunes a reflection of their choices, the condition of their lives a measure of the responsibility they have (or have not) taken for it.
>
> (Scott 2012, 9)

Building on this, I want to suggest that the concept of the individual whose conditions of emergence and realisation are defined above, has largely influenced Muslim societies living in the West, so much so that the idea of individually subscribing to the beliefs of Islam—understood as unbounded by history or socio-economic contingencies—started to take hold in France as well. Indeed, the 1980s marked a decisive turn in this transition in the context of the onslaught of neoliberalism, the fragmentation and weakening of the anti-racist movement in France, and the rise of political Islam notably in Algeria which strongly impacted the development of Islamic thinking in the French banlieues. At this point, a merciless battle began between the French republican power and the Muslim civil society, of which the first veil affair of 1989 would be the starting point.[8] Later, in 1994, an unprecedented crackdown was wrought by Charles Pasqua, then Minister of the Interior. He stepped up the rhetoric and repression against "Islamism" even further, using it as an opportunity to strengthen his security management of immigration. This repression culminated with the shooting of Khaled Kelkal in 1995.[9]

Out of this confrontation, a historical compromise was found between two antagonistic forces: I will call this the "Tariq Ramadan compromise." It represents a tacit agreement over the status quo between the Muslim community looking for justice and self-determination and the French authorities afraid of new forms of (potentially revolutionary) contestations. This compromise—and it was a compromise rather than a form of compromising since it came out of a very uneven power struggle—can be summarised by the phrase: civic Islam (in French: *islam citoyen*). Under the idea of civic Islam, Muslims can aspire and even claim individual recognition but not collective and autonomous organising. The Muslim intellectual Tariq Ramadan was the man who theorised and embodied this compromise at the end of the 90s. From then on, we see new expressions appear in public discourse that laud this compromise like "spiritual journey," "choosing one's faith," "choosing to wear the veil," "building inclusive mosques." Quranic verses that valorise individualism are also emphasised such as "religion imposes no constraints," which is used at every opportunity by reformists and modernists.

Added to the pressure exercised by the state is the one exercised by the political forces engaged in the struggle against Islamophobia, most of which will only support Muslim women on the condition that they claim being autonomous subjects free and unshackled from their community's tutelage.

I argue that the Muslims who adhere to this vision of liberal Islam mark a historical rupture with the past and with a social practice that did not foreground these cumbersome notions of free will and individual freedom as defined by liberal democracies. Yet, it is on this minefield that a naive and angelic vision of the liberation of Muslim women will flourish, championed notably by Tariq Ramadan (2004).[10]

The conditional solidarity of white progressives and existential dead end

Confronted with state racism and Islamophobia post 9/11, the French anti-racist left splits between an Islamophobic majority and a minority opposed to the March 2004 Act that excludes veiled women from public schools. While the Eurocentric white left mobilised secular, feminist, and progressist arguments to justify its support for the bill, the minority that opposed it deployed a "strategic" rhetoric to convince the white majority to join forces against Islamophobia. They mobilised arguments that build on the ideals of individual freedom, progressivism, and women's choice over their bodies. However, this rationale merely extends colonial and assimilationist frameworks, which uproot veiled women from their history and rob them of their social and community sense of belonging, leaving them to fend for themselves in a world of social and economic liberalism. Such discourse is naturally articulated with "the Tariq Ramadan compromise" mentioned above, which will provide the ideal ideological corpus and formulas to serve this strategy. Hence emerges a specific discourse that can be articulated like this: wearing the veil is a decision I took through following my own spiritual journey.

In essence, this discourse will be mobilised by Muslim women protesting Islamophobic laws when trying to justify their choice and to get it accepted by their white allies. While faith can be a personal and intimate affair, and something that is not for me to judge, social behaviours are a different matter. It is true that the question of wearing the veil for women in Muslim lands and in Muslim contexts has fluctuated through time and space. In some contexts it was abandoned, while in others it was strongly mobilised. However, the veil is indeed a significant female attribute of monotheistic religions, including Islam. So, in "safe" spaces, Muslim women are inclined to declare that they wear it as a "sign of submission to Allah," out of "religious conviction," or more prosaically, "because this is what we do." Yet, this type of discourse is clearly and completely inaudible in the French context and has failed to convince people to join the fight against Islamophobia. It is not surprising then that under social pressure in general and (political, academic, and media) pressure from the left in particular, we see new motifs emerge that conveniently fit the French secular psychology and leave no doubt as to the complete private nature of that choice. The veil as a private affair—as if the community has no influence and no say in this decision. This obviously does not hold up to any kind of serious analysis. Taking a different example, no one would think that a man in my family made me put on lipstick in the morning. On paper, it is a personal decision. Yet, we know that this choice is determined by social structures that precede me. This choice was made for me by society long before I came along. In other words, I am already "on track." I can slow down, accelerate but my options are very limited. The same is true for the veil. The fact that it appeared as a "problem" in the 80s in France is no coincidence. It is the result of a long history of resistance and self-assertion, starting with the Muslim Brotherhood in Egypt in the 30s, the rapid rise of political Islam expressed in the Iranian Islamic revolution, and followed by the failure of pan-Arabism.

In France, the veil is a manifestation of the ongoing (mostly cultural, and sometimes political) resistance of Muslims, a community far too long despised and crushed by state power, and eager to rediscover and assert its own points of reference and history. Even though Muslim women may not claim or recognise themselves in political Islam, their actions are nevertheless determined by these larger social processes and the powerful momentum they created; as well as the existential willpower of the *Oumma*.[11] This is one of the main blind spots of Islamic feminism. Pressed to demonstrate their full emancipation from the Muslim patriarchy, they come to invent an idealised version of Islam removed from its (sometimes conflicted) social, historical, cultural, and emotional context and circumstances. What is worse, they overlook the heaviest burdens carried by Muslim women in most places, from the social ghettos in the West to the working classes in the South, where they struggle in socio-economic conditions that Islamic feminism hardly problematises. Indeed, as Fatima Khemilat asserted at a feminist conference in Montreal,[12] in Western countries, sociologists have demonstrated that overall, non-white women from lower socio-economic backgrounds do not recognise themselves in feminism: firstly, because feminism has been weaponised as a civilisational tool; secondly because it is perceived as a way of demeaning non-white masculinities that are already under attack from white supremacy; and lastly, because they have other priorities considering their precarious situations. In this context, non-white women tend to more readily adopt strategies that do not harm their communities and turn away from feminism. On the other hand, it is well established that women who have moved up the social ladder, and those who access more politicised milieus like universities, or who are employed as professionals, are naturally more open and enthusiastic about feminism. Zahra Ali makes no secret of this:

> It is true that the theory of what is commonly known as Islamic feminism, where the terms feminism and Islam are theorized conjointly, primarily comes from intellectual and university settings which are elitist and restricted to an informed audience. Those behind this movement are primarily women intellectuals and social sciences researchers with a Muslim background, as well as Muslim women activists....
>
> (Ali 2020, 22)

Consequently, Islamic feminist scholars in Europe find themselves at an ideological dead end and inevitably torn. As Muslim feminists who wear the veil, they are confronted with the intolerance of the white middle and upper classes for whom the opposition between Islam and feminism is a given. While campaigning as feminists wearing the veil and trying to convince others that feminism and Islam are compatible, they have to face the hostility created by the civilisational narrative of white progressivism. In the end and because of the extreme polarisation of the debate, this often becomes an untenable position. While we can say that the unveiling of certain figures of Islamic feminism is veritably a choice, it is important to add that such choice above all emerges from within the white political majority, which has more to offer in terms of privileges and access than the poor from immigrant neighbourhoods.

Surviving like Scheherazade

As Talal Asad (Indigenes de la Republique 2015) fittingly reminds us, "we must move away from the modern idea of religious belief as something 'strictly private'."

In the chapter entitled "Allahu Akbar" of my book *Whites, Jews and Us: Toward a politics of revolutionary love*, I wrote:

> May God protect us from the word 'I.' Staying true to this adage, migrants have tried their best to preserve its deeper meaning against the grain of a dominant French perspective that glorifies the liberal, consumerist, *jouisseur* 'I'. This 'I' that drives market forces and trumps all the 'we', starting with the postcolonial 'we' which is conveniently designated as tribal. Unlike the white, bourgeois, arrogant, and cynical elites of this country, migrants share the experience of white working-class people. They know it intimately. They know how they have been stripped of God and of communism, deprived of social mobility and horizon, and disarmed to be fed to capitalism. They have seen many times the painful look of people who have witnessed the breakdown of their families, solidarities and hope. A look where they could perhaps also read this sad confession: 'at least you still have something to hold on to'.
>
> (Bouteldja, 2016, 131)

In other words, and to come back to the case of veiled women, their escape route lies in refusing to become individuals abstracted from the collective, people with no attachment to community and no history. To paraphrase the *ubuntu* proverb, "I am because we are," I would say: I am Muslim because we are Muslims.

To me, it is the survival of the Muslim community that is at stake: we have to reject the idea of liberal Islam at all costs. If we adhere to it, we will most certainly become criminalised and persecuted for simply transmitting tradition and ethics, since the liberal view does not tolerate the coercion of the individual except when it comes from the market or the state. Will we be able, as has been the case for centuries, to encourage pre-teens to observe Ramadan? The answer will be no, since the notion of "consent" is gradually becoming a cornerstone. As such, the transmission of a whole culture and its values is put in jeopardy, and it may lead the way towards the de-Islamisation of Muslims, something white powerbrokers have been wanting and planning for some time. This process, which is essentially a process of assimilation, is not unlike what happened to Jewish communities in France. We know the famous phrase from Count de Clermont-Tonnerre at the dawn of the French Revolution: "*The Jews should be denied everything as a nation, but granted everything as individuals. They must be citizens.*"[13]

Let us be clear. I am not questioning the rule of law. What I am challenging is this abstract individual who would self-realise outside of, and indifferent to, social, economic, and political forces. Because saying "*We must refuse everything to the Jews as a nation*" means denying their status as an oppressed group, denying them their right to resist as a group and therefore it means hastening their demise because it dissolves the link that unites them; indeed, it also means atomising them and making them more vulnerable to state antisemitism. De-Islamisation will not make us disappear as oppressed Muslim subjects (because social races do exist); instead, it reflects ongoing efforts and desires to separate us from each other, to push us into the open arms of the market.

And yet, this is the journey we have embarked on collectively: "The desire to become better integrated in French society has led some of us to choose secular humanism and as a consequence, reject religious dogma." This is what sociologist Houssame Bentabet observes in an interview with Hassina Mechai (2020), published in the online media *[Ehko] Média de [dé]construction* and he adds:

> This momentous change among the Muslim community announces a new reality, that of a different relationship to Islam: It means the individualisation of beliefs that places the human and its absolute freedom at the heart of Muslims' aspirations and longing for meaning.

While Bentabet seems to support this idea, it is nonetheless laden with aporias:

- First, because it places absolute freedom in contradiction with faith which is nothing more than a modern and Western idea.
- Second, because this version of liberal Islam is portrayed as a desirable horizon. But lest we forget that one transcendence tends to replace another. If Westerners, and the French in particular, have banished God from their social imaginaries, the transcendence of the Market, of the Nation, or of the Republic are felt and lived as dogmas, replacing God. Is one truly free if one believes religiously in money or in the Republic?
- Last, because this ideal ignores what Wael Hallaq aptly describes in the following Jadaliyya interview with Hasan Azad (2014):

> the Islamic so-called "legal" and intellectual traditions have repeatedly, and throughout many centuries, faced one of the most formidable questions that human societies have had to deal with for millennia; that is, the extent of moral responsibility to which the natural individual can and should bear. In every case, the Muslim jurists and their fellow ("non-legal") intellectuals, remained committed to a view that bars the waiving of moral responsibility from the individual. If the individual is the bearer of ultimate responsibility for living life, he or she must bear the onus of consequences.

Surely, this is the most political definition of an individual there is: in the Islamic tradition, the individual *does* exist, but first and foremost because he/she is responsible.

Many of us carry within us a Scheherazade, this heroine from *One Thousand and One Nights* who survived her programmed death thanks to her remarkable intelligence; and therefore, we know what is good for us and for others. We thus very consciously admit (even in a silent way) that *our bodies are not our choices*. Many will reach the hasty conclusion that this lack of ownership (more liberal vocabulary!) is a capitulation to Muslim patriarchy. But it is much more complicated than that. First, because our bodies belong to our history, to our mothers and fathers. Inspired by the Indigenous women from Mexico I heard at a conference,[14] we must be proud to claim: "our bodies do not belong to us, they belong to our community." Second, because this choice opens up the possibility of a real negotiation. While white non-decolonial feminists have mostly seen the veil as a symbol of submission to Muslim patriarchy, it represents a subtle yet powerful and radical critique of white patriarchy whose long colonial history is intertwined with the conquest and submission of Indigenous women (the veil says: my body does not belong to you); the veil is a critique of the commodification of bodies (it says: my body obeys and submits to Allah, not to the God of money); and it is a successful attempt for women to liberate themselves from the yoke of Muslim masculinity, since by making a pact with it (in refusing to break away from the community, and accepting the gender binary), women can freely move around, study and gain autonomy. This is the deeper meaning behind saying "my body does not belong to me"—the part stands in solidarity with the whole, and the whole with the part. In a context largely shaped by a form of colonial counter-revolution and the hegemony of liberal forces on a global scale, it is urgent for Islamic feminism to break from Western discursive frameworks, the binary structures of liberalism, and its individualism and mythologies.

Indeed, if we needed to destroy once and for all the myth of the benevolent and women-friendly West, it would be useful to compare the fate of Scheherazade in *One Thousand and One Nights* with her fate in Edgar Allan Poe's *The Thousand-and-Second Tale of Scheherazade*. While the "oriental" Scheherazade survives (in an environment that is harsh for women) because of her intelligence and guile (qualities that the king Shahryar recognises himself), in Poe's account, she is strangled at the end of her thousand-and-second night. While the grand Scheherazade survived her tyrannical husband, as well as the many re-writes and reinventions of the tale that reimagined and sublimed this exceptional character, she is cynically assassinated by a US writer: imperialist masculinity at its finest! And what else to expect? The civilisation that hunted and burnt witches could not possibly spare the life of this Machiavellian creature that these soft, cowardly Arabs had spared.

Notes

1 The text was originally written in French and translated into English by Esther Alloun.
2 The French law on secularity and conspicuous religious symbols in schools passed in parliament and was signed into law by President Jacques Chirac on 15 March 2004. It is also known as the headscarf or hijab ban.
3 Lallab Magazine: https://www.lallab.org/leila-alaouf-feministe-et-heureuse/
4 Bladi. Info – Forums: https://www.bladi.info/threads/theologienne-marocaine-asma-lemrabet.523156/
5 I use the word “religion” as a convenient shorthand but knowing that the word connotes a Eurocentric ambiguity which places Islam within a Christian tradition, while in reality Islam is a “Din”, that is a tradition with a widely encompassing view of life that does not separate faith from politics, or temporal and timeless matters.
6 One of the Prophet’s wives.
7 By "political Islam" I mean any political enterprise based on a legitimation by Islam which includes a wide variety of movements sometimes antagonistic to each other.
8 The 1989 incident - Three high school teenagers at a public school in a Paris outer suburb refused to unveil in class and were excluded from school.
9 Khaled Kelkal was shot dead on September 29, 1995, in France. He is a terrorist of Algerian origin, recognized as being responsible for the wave of attacks committed in France in the summer of 1995.
10 see especially chapter “The birth of Muslim feminism”.
11 Community of believers in Islam.
12 7th International Congress of Feminist Research in the Francophonie, Montreal, August 26, 2015.
13 Count Stanislas de Clermont Tonnerre: https://www.encyclopedia.com/religion/encyclopedias-almanacs-transcripts-and-maps/clermont-tonnerre-count-stanislas-dedeg
14 National Autonomous University of Mexico (UNAM), March 21, 2018.

References

Adlbi Sibai, Sirin. 2016. *La cárcel del feminismo – Hacia un pensamiento islámico decolonial*. Madrid: Akal.

Ali, Zahra, ed. 2020. *Féminismes islamiques*. Paris: La Fabrique.

Azad, Hasan. 2014. Muslims and the path of Intellectual Slavery: An Interview with Wael Hallaq (Part Two), *Jadaliyya*. https://www.jadaliyya.com/Details/30782

Badran, Margot. 2020. Féminisme islamique : qu’est-ce à dire ?. In: Zahra Ali (ed.), *Féminismes islamiques* (pp. 53–68). Paris: La Fabrique Éditions. https://doi.org/10.3917/lafab.ali.2020.01.0053

Barlas, Asma. 2020. Femmes musulmanes et oppression: lire la libération à partir du Coran. In: Zahra Ali (ed.), *Féminismes islamiques* (pp. 85–110). Paris: La Fabrique Éditions. https://doi.org/10.3917/lafab.ali.2020.01.0085

Bouteldja, Houria. 2016. *Whites, Jews, and Us*, South Pasadena: Semiotext.
de Beauvoir, Simone. 1953. *The Second Sex*. New York: Knopf.
Grosfoguel, Ramón. 2013. Hay que tomarse en serio el pensamiento crítico de los colonizados en toda su complejidad, entrevista realizada por Luis Martínez Andrade. *Revista Metapolítica* 83(17): 32–47.
Indigenes de la Republique. 2015. *Les musulmans ont-ils leur place en Occident? Entretien avec Talal Asad*: http://indigenes-republique.fr/les-musulmans-ont-ils-leur-place-en-occident-entretien-avec-talal-asad-2/
Khiari, Sadri. 2009. *La contre-révolution coloniale en France: De de Gaulle à Sarkozy*. Paris: La Fabrique Éditions.
Lamrabet, Asma. 2020. Entre refus de l'essentialisme et réforme radicale de la pensée musulmane. In: Zahra Ali (ed.), *Féminismes Islamiques* (pp. 69–84). Paris: La Fabrique Éditions. https://doi.org/10.3917/lafab.ali.2020.01.0069
Mechai, Hassina. 2020. En France, la tendance est à "l'individualisation du croire chez des musulmans", Interview with Houssame Bentabet. *[Ehko] Média de [dé] construction*.https://ehko.info/en-france-la-tendance-est-a-lindividualisation-du-croire-chez-des-musulmans/
Mir-Hosseini, Ziba. 2020. Le projet inachevé: la quête d'égalité des femmes musulmanes en Iran. In: Zahra Ali (ed.), *Féminismes islamiques* (pp. 123–150). Paris: La Fabrique Éditions. https://doi.org/10.3917/lafab.ali.2020.01.0123
Mormin-Chauvac, Léa. 2019. Interview. Nadia Tazi: "Le succès des islamistes repose moins sur la religion que sur un regain de virilité" *Libération*. https://www.liberation.fr/debats/2019/02/05/nadia-tazi-le-succes-des-islamistes-repose-moins-sur-la-religion-que-sur-un-regain-de-virilite_1707595?fbclid=IwAR3FOqt5AvhY5C6YNpT1VHnV6P3yBEpcmjZ4Rb9wFRgPjn1CNZY7Vf1M6eE
Ramadan, Tariq 2004. Western Muslims and the future Future of Islam.
Scott, Joan W. 2012. Emancipation and Equality: A Critical Genealogy, Utrecht University Repository: https://dspace.library.uu.nl/handle/1874/274997

2 Racialisation in a "raceless" nation

Muslims navigating Islamophobia in Denmark's everyday life

Amani Hassani

Introduction

Focusing on experiences of racialisation, this chapter draws our attention to the power dynamics at play that uphold the structures of dominance. Any attention to this from the perspective of Muslims' experiences, inadvertently becomes an attention to the creative ways these citizens contest, negotiate, and navigate through such power dynamics. Agency as a *capacity for action* is not always rooted in overt resistance, but sometimes more implicit—maybe even docile—ways of ensuring one's right to express one's Muslimness (Mahmood 2011). With this emphasis, this chapter unpacks the experiences of racialisation in Denmark, a colourblind progressive liberal society where the academic scholarship surrounding Muslims has too often focused on their "foreignness"; i.e. their religious organisation, integration processes, cultural differences, transnational ties, etc. (Rytter 2019). This has somewhat skewed the academic conversations—possibly even helped reproduce the political reification of Muslims as "Other" in Denmark—with little attention given to structural power dynamics through which Muslim citizens are surveilled and their social lives scrutinised and intervened in. Based on ethnographic fieldwork conducted among young Muslims in Copenhagen, this chapter demonstrates how they experience racialisation and Islamophobia in everyday life. I use the concept of Islamophobia throughout this chapter, not merely as religious discrimination, but to emphasise how structures of power racialise Muslims as a quintessential Other—diametrically opposite to (and incompatible with) "Western civilisation." In fact, an ethnonationalist shift in Danish political rhetoric since the early 2000s simultaneous with the global War on Terror, has meant that the racialisation of Muslims has become politically pronounced, and Islamophobic legislation has been introduced under the guise of protecting Danish liberal democracy from a "Muslim threat." In this chapter, I argue that this type of structural Islamophobia is promoted through political discourse and mainstreaming of anti-Muslim racism (Fekete 2018), which inadvertently trickles down to everyday social and spatial interactions.

DOI: 10.4324/9781003293323-3

Racialisation in a nation without race

The dominant national narrative in Denmark emphasises its progressiveness, liberalism, and welfare ideals. It is a political model for the world to admire (Jensen 2018). This is a narrative founded on the premise of a benevolent civilising coloniser, as Lars Jensen demonstrates in his historical analysis of Danish imperial past. This narrative translates into contemporary political discourse, particularly directed at migrants and descendants of migrants from the Global South. The national self-image is of its own goodness—despite its colonial past (and present). The Dane was a different sort of coloniser: "good," "humane," and "benevolent" (Jensen 2018, 67–68). This chapter disrupts this image, not by dismantling the hegemonic narrative but rather by interrogating the repercussions of the narrative on Denmark's racialised Others, particularly Muslims. By focusing on Muslim youth coming of age in post-9/11 Denmark, I want to challenge the idea of Denmark as a post-racial state, both in political rhetoric and popular national imagination. Instead, I suggest a way to understand Danish political structures as a social system in which a racialised hierarchy is infused within both the political and social structures while simultaneously neglecting the significance of racism. Bonilla-Silva's conceptualisation of racialised social systems is useful to think with in this regard, as an expression of how economic, political, social, and ideological structures depend on a racialisation of people (Bonilla-Silva 1997, 469). Bonilla-Silva draws on Omi and Winant's (Omi and Winant 2014 [1986]) theory of the formation of race as a social phenomenon founded in social structures. In other words, rather than discussing race in biological terms, Omi and Winant argue that racialisation is a social process closely connected to social, economic, and political institutions. Quijano (2000) situates this process of racialisation within an analysis of global power dynamics that is closely tied to a European colonial past and an enduring Eurocentrism. In this sense, racialised hierarchies need to be understood as a legacy of colonialism, which have become a stable feature of global (and in this case, national) processes of power (Quijano 2000). By understanding racial hierarchies within the frames of European colonial past, we can understand contemporary expressions of racism, where "race" is never explicitly acknowledged yet remains present in public consciousness (Younis 2021). Rather, "cultural racism" is a way of culturally inferiorising, in this case, the (Muslim) Other. More specifically, Islamophobia becomes a type of cultural racism through which the Muslim becomes the inferior, uncivilised, potentially dangerous barbarian as opposed to the civilised (white) European.

There has been limited attention in academic discourse to understand how racialised social systems exist outside the American or the British context (Gullestad 2002). This is, however, changing, as topics of migration and border control and the integration of migrants and their descendants within the Global North is becoming a political issue with strong racist undertones. This demands a reconceptualisation of what race and racism means within these societies (Hansen and Suárez-Krabbe 2018; Hervik 2019; Suárez-Krabbe and Lindberg 2019; Hassani 2020).

Paul Gilroy (2013) established a long time ago how nationalism and racism are interconnected concepts. One feeds off the other in a racialised social system. And yet in Denmark, nationalism is hailed as the protection of liberal values while racism is tabooed as non-existent. Thus, the Danish political and social system perceives itself as colourblind; race is not and has never been a problem in any serious way (Danbolt and Myong 2019). This means that racialising discourse on the Muslim Other promotes culturalist explanations of "essential foreignness" that need "integrating" into progressive liberal values (Omi and Winant 2014). Nevertheless, there's an erasure of race in Denmark—not in actual effect, but merely in public discourse. One can rarely talk of racism—whether structural or individual—although it permeates through society at the political, economic, and social levels.

By understanding racism as a by-product of racialisation within a racialised social system, the emphasis is on how it rationalises the structures of racialised inequality as well as racist encounters in everyday life (Bonilla-Silva 1997, 474). In the Danish context, the racialisation of the Muslim Other becomes commonsensical (cf. Omi and Winant 2014), for example: Muslims are from non-Western countries, they are misogynistic, they are backwards, and they are "guests" in "our nation" (Hervik 2004; Jaffe-Walter 2016). What is important to draw from this is how the racialisation of Others—even in nations such as Denmark, where the population is over 95% white—becomes fundamental to upholding the power dynamics within society.

The civilised nation and the "threatening" Other

In her book on Dutch racism *White Innocence*, Gloria Wekker (2016) introduces us to a conceptual framework to understand how racism plays out in societies that have neither dealt with their colonial past nor the entrenched racialised structures that have resulted from this past. Like Denmark, the Netherlands perceives itself as a benevolent small and just welfare society. With this self-perception, the very notion of racialisation and racism as integral parts of the social structures beyond an individualised expression cannot be entertained (ibid.). In Denmark, there are laws to prevent discrimination and hate speech against people's race, ethnicity, religion, or sexuality. Nevertheless, the concepts of race and racism are often socially and politically dismissed as not relevant to the socio-historical context (see e.g. Hansen & Suárez-Krabbe 2018). Instead, the idea of ethnicity and other cultural signifiers are emphasised to highlight foreignness and non-belonging in the guise of "inclusion" through integration, which is often meant as assimilation into hegemonic white culture.

Wekker unpacks how race and racialisation within the dominant discourses—in national politics, media, and popular entertainment—goes unchecked because of the national(ist) imagination of the country's progressive liberalism. This progressive liberalism is perceived to have eradicated racial hierarchies and, by extension, racism. Nadia Fadil (2010), focusing on Flanders in Belgium (the Dutch speaking region), connects this idea of the benevolent welfare state to the notion of Islamophobia. Similar to Wekker, Fadil highlights how in the Flanders discourse there is a racialised differentiation between *autochton* (native Dutch) and *allochton* (often racialised to non-Western people of colour). In Denmark, it is the difference between Danish and non-Western immigrants and descendants (an official category to differentiate non-white people from white Danes), who in the past 20+ years have been racialised as Muslims (Yilmaz 2016).

The similarity in these cases is the national imagination of liberal values, which are inadvertently threatened by "Muslim" illiberalness. Fadil explains this point, describing how Muslims' lives "which fall outside the liberal spectrum, are seen as 'barbaric', incapable of similar liberal values, and even potentially 'threatening' to one's own liberal lifestyle" (2010:249). This dichotomy between the civilised Nation and the threatening Other is not a new phenomenon but has fuelled much of national(ist) rhetoric in the guise of national exceptionalism. While none of the popular images of the nation within these countries would admit a racist past, not to mention a racist present, there is an image of national(ist) exceptionalism based on its progressive welfare. If we understand nationalism as a boundary-making process in which people perceived as Others are excluded, the argument introduced by Paul Gilroy over three decades ago becomes essential to understanding how nationalism within these progressive liberal societies is deeply interconnected with a racialised social system (Gilroy 2013 [1987]).

Making difference palatable

So far in this chapter, I have presented a conceptual framework to understand how Denmark perceives its racialised Others with a particular focus on the Muslim as the quintessential Other. In the following analysis, I draw on ethnographic fieldwork conducted among young Danish Muslims, who are highly educated and socially mobile, and thus able to challenge the dominant racist stereotype of the socioeconomically and culturally inferior Muslim foreigner. Through interviews and urban walks, I sought to understand how these youth came of age in the years following 9/11. How have they experienced the growing political nationalism and explicit Islamophobia, and how have these political tendencies influenced their everyday lives and interactions with white Danes?

For the past two decades, if not more, the hijab and niqab have often become the centre of political attention when discussing overt religious symbols in Denmark with political attempts to regulate Muslim women's dress (cf. the current ban on (Muslim women's) face covering in all public spaces in Denmark). As such, the hijab was a prevalent symbol in my female interlocutors' narratives, whether they wore it or not. My male interlocutors did not face the same kind of overt reification of their Muslimness based on their dress. They did, however, face a different set of challenges, which meant they often put a lot of attention into presenting themselves as *Danish* Muslims as opposed to "non-Western" immigrant Muslims. Yet because of how Muslim men are racialised in Danish popular discourse, they often felt a need to pre-emptively deal with the idea that they were not just seen as young men but rather as brown/black males = Muslim = immigrant = criminal/radical/violent. These stereotypes of "violent Muslim men" meant that the young men I met invested time and effort in becoming palatable to white gazes. Idris describes this when he explained how he engaged with his high school teacher:

> I make them understand that immigrants aren't criminal or violent. For instance, my English teacher, who used to be very bigoted, actually told me that I changed her mind.
>
> (Idris, interview[1])

Rather than blame his teacher for her racism and bigotry, he emphasised his own ability to perform in such a way that reduced the teacher's racism rather than resist it. This demonstrates the complex challenges that these young men face trying to navigate a system that vilifies them *a priori*. Khalid, another young man, explained this further when we were talking about how his employer initially didn't like him—presumably because of his Muslim and ethnic background:

> [...] You have to step up when you need to. You also have to make an effort. You have to build relationships, and that's what I did with my [supervisor]. Initially, she didn't like me and didn't want to hire me [after my internship], but I made a good impression and was open with her, when we spoke about our private life [...], so they also know that you're a person. [...] The [supervisor] definitely had her prejudice that had to be taken down slowly.
>
> Isn't it annoying having to deal with this type of prejudice?
> It's not annoying. I see it as a necessity. You're not like everyone else in Denmark, and you have to remember you're Muslim first and foremost, and you should show your best qualities, because then you can't fail—to display good manners and *akhlaq*.
>
> (Khalid, interview)

As a Muslim man, Khalid had to respond to his white female supervisor's prejudices against him to develop a better rapport with her. This meant that he had to be sensitive and take the time to answer her questions about his personal and religious life, making his lifestyle palatable to her gaze. However, he described refuting people's preconceived notions regarding him as matter-of-fact; as a *Muslim* minority in Denmark, you are not like everyone else and you *have to* represent good manners—you have to be respectable (cf. Higgenbotham 1992).

My interlocutors often seemed to have to navigate the difficult terrain between respectability and their self-chosen Muslimness, e.g. in attempting to adhere to their own sense of religiosity. For instance, Musa, a young man working in the hotel business while finishing his university degree, explained:

> Initially [...], I try to indicate that I'm Muslim. [...] So, if a colleague invites me to go to Christmas lunch or something else non-Islamic, I try to reject it in a subtle way [...]. A problem I often face because I work in a hotel ... most receptionists are young women, so there's a chance of physical contact, for example shaking hands, so I usually use the line [...] "I only shake hands with the old or ugly, and you're neither"—they take that pretty well.
>
> (Musa, interview)

The symbolic gesture of shaking hands has been a contested marker of difference between "the white Danish us and the Muslim Other". It has been interpreted by politicians as a sign of disrespect of gender equality to the point that the now former Integration Minister, Inger Støjberg, passed it into law as part of the citizenship ceremony, arguing:

> If you don't want to shake hands with the mayor [during the citizenship ceremony], it can be understood as a clear sign that you don't believe in gender equality. That's inherently why there are Muslims who don't want to shake hands with the opposite gender. So, I definitely think you can say that this is essential.
>
> (quoted in DR, translated from Danish, Olsen 2018)

It is in this context that Musa attempts to negotiate his position. The only way to make it palatable to his white female colleagues is by being tongue-in-cheek and attempting to flatter them.

This idea of making differences palatable was important to several of my male interlocutors in a different way than to the young women I spoke to. For instance, Iman explained when she decided to wear the hijab—it is difficult make the hijab completely palatable because it is a visible symbol of difference—it was a constant struggle to be reduced to the stereotype of "the oppressed Muslim woman":

> I met resistance from the surrounding society saying I should take it [the hijab] off: "Tell your father you don't have to wear it!". I often felt defensive: "I actually have a free choice. No, my parents aren't strict or have high academic requirements for me, and no I will not be forced into a marriage." It required a lot of energy to continuously try to climb out of the box they put you in, especially in high school.
>
> (Iman, interview)

I interviewed her in 2013 around the same time that the concept of "negative social control" was first introduced in relation to specifically Muslim women's sexual relationships vis-à-vis their parents' expectations. Iman's comment here demonstrates how this rhetoric of "social control" was based on a racialised understanding of Muslim men as inherently oppressive towards Muslim women. Meanwhile, reality for Iman, who was 21 years old at the time, was that she chose not to disclose to her colleagues or university peers that she was married. She worried they would potentially put her in "a box" as she described it. So, although she was academically and professionally successful, the social consequences of the government's Islamophobic policies and rhetoric meant that she had to censor herself and her personal life to the white gaze.

Structural privilege in spatial interactions—observations

The young Muslims I have presented so far find various ways of negotiating and challenging their positions vis-à-vis their racialisation in everyday spaces and social interactions. In this section, I want to present an experience I had—as a self-ascribed *and* racialised Muslim—in the early days of my ethnographic fieldwork in Copenhagen. In the fall 2013, the newly built central mosque in Copenhagen had an open-door event for *the Night of Culture*—a yearly cultural event facilitated by Copenhagen municipality giving the public an opportunity to engage with local organisations, businesses, institutions, etc. after normal opening hours. I decided to go and take in all the impressions of the new building, its architecture and the people attending. The mosque had not officially opened and there were still parts of it under construction. Nevertheless, it was clear that it was going to be an impressive building with aesthetically pleasing fusion design elements—in the words of the guide: linking the Danish colour palate with Oriental geometric designs and Quranic decorative scripture. On the tour of the mosque, the guide directed my attention to how the organisation behind the mosque was attempting to be inclusive and open to the wider Danish (non-Muslim) society by offering spaces for dialogue and incorporating architectural elements that fit with the Copenhagen urbanscape. They even went as far as including a mural of Copenhagen's skyline which included the mosque; literally painting Muslims' presence into the fabric of Copenhagen's landmarks.

Immediately following the tour, I met two young white non-Muslim women outside the mosque building. They had also been on a guided tour of the mosque that evening, and we were all three processing our impressions. Like me, they also noticed the inclusive narrative throughout the tour and were positive about their experience. One of the women, however, shared that although everything seemed impressive, she was still a bit concerned about the fact that Qatar was the main financial contributor. Who is going to decide who is going to be the imam, she argued, and how will this be displayed? After all, they're Wahhabis in Qatar, she proclaimed. Attempting to ease her worries, I suggested that Qatar was more aware of Western culture. Also, I eagerly continued, if they end up deciding who the imam is, he would not be able to relate to the Danish context anyways. I finally retorted that the mosque activities must be in Danish to be in keeping with the inclusive tone the mosque presented at the event. She agreed, but again emphasised that she hoped there would be a way to recruit local people to the mosque. Even with this suggestion she seemed sceptical, because how would they choose a person to be the imam? If only Denmark could get it together and introduce an Imam-degree—like Christian theologians—we wouldn't be in this predicament, she insisted. I responded—a bit defensive, perhaps even protective—that I did not see any need for formalising an Imam-degree, but would rather give younger people the opportunity to understand and proclaim their beliefs within the context they live in.

Changing the subject, I asked them if they were actually going to use the facilities in the community centre linked to the mosque. The same young woman responded that she would, especially the restaurant, which she would consider bringing her mother to and challenge some of her prejudices. She told me that she, herself, was not scared of religion, she actually felt very comfortable talking and socialising with Muslims. It was more the older generation—her mother's generation—who were scared of the unknown. She didn't think religion played a role when she socialised with people, and that is also how she thought Muslims in Copenhagen experienced it. You socialise based on similar interests, being from Nørrebro (the culturally diverse area of Copenhagen) and living in this part of the city.[2]

This interaction seems underwhelming; at least it initially was to me. So much so I actually forgot about it while I was processing my fieldwork notes. The ideas of racialisation and space and the interplay between them where not my main objective when I first travelled to Copenhagen to do the first half of a comparative ethnography of young Muslims in progressive liberal societies. However, as an explorative ethnographer, I knew there was something significant about this seemingly unextraordinary interaction with these two young women. It was perhaps less about their responses to my questions that made me describe this interaction in my field journal with great detail. It was more about what I felt as I engaged with them; defensive, insecure, perhaps even out-of-place, or more precisely *put* out-of-place.

Returning back to this journal entry years later, its significance became apparent as I started to unpack the responses of my Muslim interlocutors, their feelings of apprehension in being in "white" spaces (Anderson 2015) or even making claims to spaces where they are perceived to be foreign (cf. the idea of being 'space invaders', Puwar 2004). In this vignette, however, I was not standing in a "white" space: it was a mosque, and although I did not know the mosque or the organisation that runs it, I felt this was at least a "Muslim" space, perhaps even a *safe* space where the only Danes there were either Muslims themselves or felt an allyship with Muslims in their continued struggle to achieve recognition as part of the Danish national image. The conversation with these two young women rattled me, perhaps because I realised that on this *Night of Culture*, it was not about developing a relationship between the mosque and local residents. Rather, I felt it was a way of *consuming* Muslim "culture" and narratives, and by extension *assessing* the Muslim Other (represented by the mosque) based on whatever narrative it presented: "Do they *really* see themselves as part of Denmark? Are they really loyal to *our* nation?" In my interaction with these two young women, the space I had thought of as *safe* had become "white," as in "hegemonically white." I wasn't just talking to two young women; I was defending the Muslim Other to the Danish public.

In retrospect it was unnerving, not that I hadn't experienced it before. I've been to many of these events where my *safe* space was turned *public* i.e. where the hegemonical white gaze invaded *my* space. And yet, this does not work quite the same way the other way around. That is not to say the young Muslims I met did not invade "white spaces" all the time, but when they did, they were often conscious of it either as a way of contesting the spatial/social exclusion or as a way of socially positioning themselves within these spaces (cf. Puwar 2004). Either way, as demonstrated in the examples in the previous section, these young Muslims trod deliberately and consciously. In contrast, I don't believe the two women in the vignette were at all conscious about the racial powerplay I experienced in our interaction. They didn't *intend* violence in their suspicions, but intentions mean little when ones' actions uphold and reproduce racialised power structures. In other words, within a racialised social system, these young women were in a position of power: they could interrogate, question, and undermine the Muslim Other within the most "Muslim" space—the mosque—in Denmark.

In the previous paragraph, we see how Idris, Khalid, Musa, and Iman all negotiated their positions with a constant consciousness of being evaluated by a white gaze; they censured themselves, they divulged their personal lives, and they flattered their white superiors, colleagues, and peers to become accepted as respectable (Muslim) Danes. In comparison, these two women did not hesitate to share their concerns and suspicions about the mosque to me, a visibly Muslim woman. Their proximity to political power *qua* their whiteness meant that they felt entitled to question the Muslim space, reproducing the ethnonationalist perspective of responsibilising the Muslim Other to demonstrate their belonging. It is in unpacking such racialised dynamics within spatial interactions that we come to appreciate how the connection between space and racialised interactions must be placed within larger social power structures.

Experiencing Danish Islamophobia

Danish Muslims experiences cannot be disconnected from the structural realities which privilege whiteness and its proximities (Gullestad 2002). Ethnonationalism as the structural backdrop in which Muslims (and other racialised minorities) have to navigate, is not only about how white people treat racialised individuals. Rather, it is a racialised social system that influence most areas of minorities' lives in Denmark. It is their consciousness of this system (although not always made explicit) which enables minorities to navigate, resist, and contest these structures in everyday interactions.

There have been decades of critical race scholarship which has demonstrated how the US operates as a racialised social system (Bonilla-Silva 1997;

Omi and Winant 2014; Anderson 2015). It is, however, more controversial to claim that Denmark is a society which works on a racialised hierarchy. The fact remains, however, that as soon as there is a public perception of who "real" Danes are, there is also a racialisation of the Other which will intrinsically affect the everyday lives and livelihoods of these people. Going back to the point on Muslims—in all their heterogeneity—as a racialised Other, we need to unpack what this actually means within a racialised social system. According to Garner and Selod (2015), racialisation is a tool for the rulers to implement on the ruled: "[i]t draws a line around all the members of the group; instigates 'group-ness', and ascribes characteristics, sometimes because of work, sometimes because of ideas of where the group comes from, what it believes in, or how it organizes itself socially and culturally" (2015, 15). The point here is that racialisation is not the same as self-ascription, i.e. Muslims feeling a sense of community based on similar religious beliefs. Rather, racialisation is implemented top-down by the power structures that perpetuate the racialised system (e.g. governments, mainstream media, entertainment industry, etc). These representations "transform the clearly culturally and phenotypically dissimilar individuals [...] into a homogenous bloc: this is the basis of the racialization of Muslims (the process), and of Islamophobia (the snapshot of outcomes of this process)" (ibid.). Islamophobia—as with racism in general—is thus a by-product of a racialised social system.

As one of the earliest and most important thinkers in American sociology, W.E.B. Du Bois (2008 [1908]) provides an important reference to understand how racialised people internalise their racialisation through a double consciousness. They "learn to read themselves through the eyes and mindsets of the majority population and regulate their behaviour accordingly in specific contexts" (Garner & Selod 2015, 17). The idea of double consciousness is woven into many of the narratives presented in this paper. While it is expressed in different ways, all the young Muslims I met had to deal with ways of being racialised as Muslims and acted accordingly: they negotiated their self-representation as a "good"/respectable minority and their position within social spaces often as a *response* to their racialisation.

With this understanding of the racialisation of Muslims as a process to *Otherise* them, our attention is drawn to the physical markers through dress and phenotypical signs, but also their practices of Muslimness, such as daily prayers, fasting, halal food, etc. In this sense, it is not only the external signs of Muslimness that are racialised, but their very practice of Islam is racialised as Other.

Racialisation of Muslims and structural expressions of Islamophobia—as a way for political powers to curb and limit Muslim agency (Sayyid and Vakil 2010)—tells us something about the social structures and how racialised power dynamics are upheld from the top, trickling down to real life consequences in social interactions as well as life chances (e.g. educational limitations, employment opportunities, housing inequalities, etc.). They create a powerful infrastructure of subordination through which Muslims in these societies must navigate (Massoumi, Mills, and Miller 2017, 13–14).

The Muslim subject and everyday life

Islamophobia—as a racism towards Muslimness or perceived Muslimness—is becoming increasingly entrenched within the political and populistic Danish discourse (Hassani 2020). The young Muslims I met in Copenhagen were adamant about *being* a part of and being *seen* as part of these cities and nations, and they resisted—sometimes explicitly, but often implicitly—attempts to sideline their belonging. However, their resistance needs to be placed within the realities they reside within: a national(ist) context which perceives them as Other/minority/non-Western which are all racialised to Muslims.

Within this frame, Saba Mahmood's (2011) conceptualisation of agency is useful to think with. Building on poststructuralist theorists' understanding of power as structural and permeating all social life (cf. Butler), Mahmood contends that agency is not merely a resistance to subordination. Rather, agency is "a capacity for action that specific relations of *subordination* create and enable" (italics in original, Mahmood 2011, 18). In this sense, Danish Muslims' agency—i.e. *capacity for action*—is directed by the structural subordination they experience, which curbs their ability to exercise the full breath of liberty that their white counterparts enjoy (e.g. dress, speech, etc).

Otherisation as a form of subordination, in this sense, does not curb all capacity for action. Nevertheless, the capacity to resist and contest the power dynamics is limited to hegemonic narratives. In other words, the young people I met had a capacity to challenge, rephrase, and reframe the presumptions of their political subordination, but they were not necessarily able to dismantle, disempower, or circumvent these structures of power.

Conclusion

Throughout this chapter, I have argued for a reconceptualisation of Denmark as a nation that is built on a racialised social system through which the Otherisation of its minorities, particularly Muslim minorities, uphold structures of power that subjugate racialised Others and privilege racialised white Danes. Being a racialised Muslim is thus not a question of religious self-ascription but a question of political subjugation that is expressed in social, political, and even spatial terms, where being or being perceived as Muslim is a catalyst for suspicion, questioning, and eeriness in public and political discourse which trickle into everyday life. I have argued that the young Muslims I met in Denmark did not always take a direct stance in resisting these associations with "threat" but rather navigated through them, figuring out ways to become palatable and respectable by their superiors, colleagues, and peers. The ways in which Muslim racialisation permeates through all levels of Danish society means that Muslims have to constantly relate to an almost omnipresent white hegemonic gaze, where even their most intimate relationships are scrutinised for signs of "backwardness" or "threat" to the civilised white progressive liberal nation.

A number of factors have without a doubt influenced young Muslims' everyday experiences and interactions with white Danes: the heightened racialised and ethnonationalist rhetoric in Danish political discourse, the global War on Terror, as well as the hostile border control of asylum seekers and migrants from the Global South, to name a few. Islamophobia in this regard is the act by political powers to (sometimes forcefully) curb Muslim political agency (Sayyid & Vakil 2010). This in turn, I would argue, has created a circumstance where Danish Muslims are required to adhere to (white) hegemonic social values to be acceptable. While my young interlocutors did not wish to give up on their religious values, they knew that some of their religious or cultural practices were "frowned upon" by white Danes. This meant, for instance, that 21-year-old Iman did not tell her peers at university that she was married, because although having an intimate partner at 21 is perfectly normal if not encouraged, having a husband at 21 as a Muslim woman is a potential sign of patriarchal oppression (especially if the woman also wears the hijab).

Why was it so important for the young people I met to be "acceptable/respectable" especially in their interactions with school or professional settings (i.e. white spaces, cf. Anderson 2015)? They came of age post 9/11, and they learnt early in life that to be a "good Muslim" in a political sense (Kundnani 2014) is to not be too vocal about your critique of society, not be too political about your social or political opinions, and not be too angry in your resistance against racism. Instead, you have to be *subtle* in your difference (cf. Musa), *change* racists' opinions by your good behaviour (cf. Idris), take down prejudice *slowly* by being open about your personal life (cf. Khalid), and not disclose aspects of your life that can put you in a *box* (cf. Iman). Meanwhile, the two young white women I met at the open evening event at the central mosque had learnt the exact opposite growing up in the same era: you have to question Muslim Others. In other words, white suspicion is necessary to supposedly protect Danish values of equality, freedom, democracy, from a Muslim Other, which is imagined to represent the quintessential opposite of and threat to these values. The power discrepancy is stark in comparing my cases of Muslim navigation in white spaces vis-à-vis white Danes entering and questioning the Muslim Other in "safe" spaces such as the mosque.

Notes

1 All interviews were conducted in Danish and quotes were subsequently translated by Hassani.
2 Field journal, translated from Danish, October 11, 2013.

References

Anderson, Elijah. 2015. 'The White Space'. *Sociology of Race and Ethnicity* 1 (1): 10–21.

Bonilla-Silva, Eduardo. 1997. 'Rethinking Racism: Toward a Structural Interpretation'. *American Sociological Review* 62(3): 465–80.
Danbolt, Mathias, and Lene Myong. 2019. 'Racial Turns and Returns: Recalibrations of Racial Exceptionalism in Danish Public Debates on Racism'. In *Racialization, Racism, and Anti-Racism in the Nordic Countries*, edited by Peter Hervik, 39–61. Approaches to Social Inequality and Difference. Cham: Springer International Publishing. https://doi.org/10.1007/978-3-319-74630-2_2
Du Bois, William Edward Burghardt. 2008. *The Souls of Black Folk*. Oxford: Oxford University Press.
Fadil, Nadia. 2010. 'Breaking the Taboo of Multiculturalism. The Belgian Left and Islam'. In *Thinking Through Islamophobia: Global Perspectives*, edited by Salman Sayyid and AbdoolKarim Vakil. London/NewYork: Hurst & Columbia University Press.
Fekete, Elizabeth. 2018. *Europe's Fault Lines: Racism and the Rise of the Right*. London; New York: Verso.
Garner, Steve, and Saher Selod. 2015. 'The Racialization of Muslims: Empirical Studies of Islamophobia'. *Critical Sociology* 41 (1): 9–19.
Gilroy, Paul. 2013. *There ain't no Black in the Union Jack*. London: Routledge.
Gullestad, Marianne. 2002. 'Invisible Fences: Egalitarianism, Nationalism and Racism'. *Journal of the Royal Anthropological Institute* 8 (1): 45–63.
Hansen, Nanna Kirstine Leets, and Julia Suárez-Krabbe. 2018. 'Introduction: Taking Racism Seriously'. *KULT. Postkolonial Temaserie* 15: 1–10.
Hassani, Amani. 2020. 'Islamophobia in Denmark: National Report 2019'. *European Islamophobia Report 2019*.
Hervik, Peter. 2004. 'The Danish Cultural World of Unbridgeable Differences'. *Ethnos* 69 (2): 247–67.
Hervik, Peter. 2019. 'Denmark's Blond Vision and the Fractal Logics of a Nation in Danger'. *Identities*, 2019: 1–17. https://doi.org/10.1080/1070289X.2019.1587905
Jaffe-Walter, Reva. 2016. *Coercive Concern: Nationalism, Liberalism, and the Schooling of Muslim Youth*. Stanford: Stanford University Press.
Jensen, Lars. 2018. *Postcolonial Denmark: Nation Narration in a Crisis Ridden Europe*. London/NewYork: Routledge.
Kundnani, Arun. 2014. *The Muslims Are Coming!: Islamophobia, Extremism, and the Domestic War on Terror*. NewYork: Verso Books.
Mahmood, Saba. 2011. *Politics of Piety: The Islamic Revival and the Feminist Subject*. Princeton: Princeton University Press.
Massoumi, Narzanin, Tom Mills, and David Miller, eds. 2017. *What Is Islamophobia?: Racism, Social Movements and the State*. London: Pluto Press.
Olsen, Theis Lange. 2018. 'Inger Støjberg Efter V-Kritik Af Håndtryk: "Ingen Er Tvunget Til at Blive Dansker"'. *DR*, 2018. https://www.dr.dk/nyheder/politik/inger-stoejberg-efter-v-kritik-af-haandtryk-ingen-er-tvunget-til-blive-dansker
Omi, M., and H. Winant. 2014. *Racial Formation in the United States*. Routledge.
Puwar, Nirmal. 2004. *Space Invaders: Race, Gender and Bodies Out of Place*.
Quijano, Anibal. 2000. 'Coloniality of Power and Eurocentrism in Latin America'. *International Sociology* 15 (2): 215–32.
Rytter, Mikkel. 2019. 'Writing Against Integration: Danish Imaginaries of Culture, Race and Belonging'. *Ethnos* 84 (4): 678–97. https://doi.org/10.1080/00141844.2018.1458745
Sayyid, Salman, and AbdoolKarim Vakil. 2010. *Thinking Through Islamophobia: Global Perspectives*. Redwood: Cinco Puntos Press.

Suárez-Krabbe, Julia, and Annika Lindberg. 2019. 'Enforcing Apartheid?: The Politics of "Intolerability" in the Danish Migration and Integration Regimes'. *Migration and Society* 2 (1): 90–97.

Wekker, Gloria. 2016. *White Innocence: Paradoxes of Colonialism and Race*. Durham: Duke University Press.

Yilmaz, Ferruh. 2016. *How the Workers Became Muslims: Immigration, Culture, and Hegemonic Transformation in Europe*. Aesthetics: University of Michigan Press.

Younis, Tarek. 2021. 'The Muddle of Institutional Racism in Mental Health'. *Sociology of Health & Illness* 43 (8): 1831–39. https://doi.org/10.1111/1467-9566.13286

3 Enriching Sami language distance education[1]

Hanna Helander, Satu-Marjut Pieski, and Pigga Keskitalo

Introduction

The Sami have endured long periods of civilisation, assimilation, and nationalism caused by the Christian Church and nation states, resulting partly in their loss of identity and language. It was not until 1970, with the establishment of a national primary school system in Nordic countries with renewed school laws, that a more inclusive system was opened for the Sami (Nyyssönen 2013). Despite the progress in the field of education and language revitalisation efforts, Sami languages are classified as endangered, and the process of language change continues (Keskitalo, Määttä, and Uusiautti 2012). This predicament requires active work to revitalise Sami languages and cultures.

The colonisation of the Sami people has left a differently nuanced history than, for example, those in the Americas, Africa, Australia, and New Zealand; however, the current Nordic countries are established, at least partly, on Sami land, and different kinds of colonial practices have been imposed on the Sami people and their culture in these countries (Lehtola 2015). In an educational context, we discuss the active measures directed at the Sami people and their cultures, entailing skewed power relations and minoritised positions since the 1600s. Colonisation has been carried out by excluding Indigenous peoples, framing them as a problem and disrespecting both their culture and knowledge (Geia and Sweet 2012, 2). As Kuokkanen (2000) stated, this has resulted in a colonised mindset. These starting points indicate the need to rethink the practices of Sami education within current contexts.

In light of this colonial history, the institutionalised education of Indigenous people needs to change. This involves indigenising educational practices, such as by applying traditional child-rearing practices in the school setting in an emancipatory way. This requires the active identification of aspects that require further development and broader collaboration to implement these developmental changes. Sami children are linguistically diverse, largely due to the pressures of assimilation. At the same time, Sami language education is challenging due to a shortage of competent teachers, updated learning materials and resources (Arola 2020). Thus, developing teachers' competency and elaborating on ways to work in diverse contexts framed by the reality of Indigenous education are required.

DOI: 10.4324/9781003293323-4

The Sami people live in four countries: Mid and North Sweden, Mid and North Norway, North Finland, and the Kola Peninsula of the Russian Federation (Map). It is estimated that around 100,000–150,000 Sami reside in these countries (Brzozowski 2021). Of these, around 30,000–40,000 persons speak one of the nine Sami languages, which, according to UNESCO Atlas of endangered languages, are endangered in various ways (Salminen 2010).

This chapter focusses on Sami language distance education in Finland. About 10,000 Sami live in Finland, of which about 3,000 speak one of the three Sami languages spoken in the country. Northern Sami has an estimated 2,000 speakers in Finland (the total number of speakers in Norway, Sweden and Finland is estimated to be 20,000–30,000, making it a seriously endangered language). Severely endangered Inari Sami and Skolt Sami are spoken only in Finland, with about 400–500 Inari Sami speakers and 300 Skolt Sami speakers (Pasanen 2016).

Currently, about 75 per cent of Sami children live outside the Sami homeland, which covers the Utsjoki, Inari, Sodankylä (northern region), and

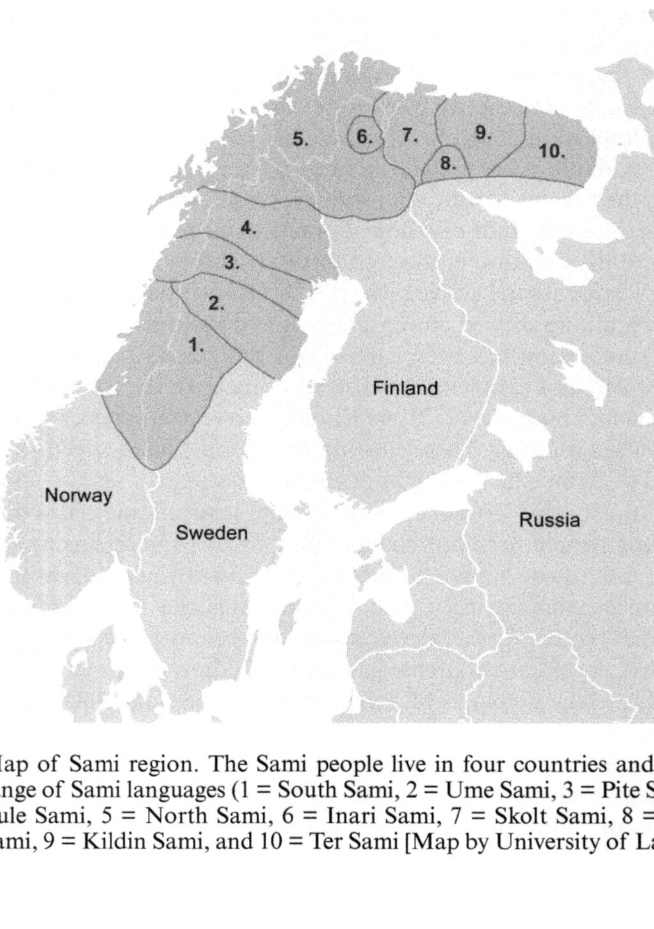

Map 3.1 Map of Sami region. The Sami people live in four countries and speak a range of Sami languages (1 = South Sami, 2 = Ume Sami, 3 = Pite Sami, 4 = Lule Sami, 5 = North Sami, 6 = Inari Sami, 7 = Skolt Sami, 8 = Akkala Sami, 9 = Kildin Sami, and 10 = Ter Sami [Map by University of Lapland]).

Enontekiö municipalities. Sami language education is secured under the Basic Education Act.[2] Outside the Sami homeland, a special regulation by the Ministry of Education has been in force since 2003 covering the costs of a supplementary two hours of Sami education in schools[2] that occurs in many ways—teaching is organised as contact lessons in the largest cities, while another option is to participate in distance-learning lessons.

In Finland, a pilot project on distance education in Sami languages (henceforth, the Pilot project) was launched in 2018 to facilitate accessibility to Sami language education.[3] In addition, about 100 children participated in contact teaching outside the Sami homeland. Given that about 2,000 children live outside the Sami homeland, it is noteworthy that only 10 per cent of them attend Sami language education (Lehtola and Ruotsala 2017).

This chapter examines the cooperation between the international ADVOST[4] research project and the Pilot project on distance education in Sami languages. One form of Sami language education in Finland is explored, providing an example of a pedagogical innovation study conducted in a distance-learning environment based on the idea of benefiting teachers' expertise by changing teaching practices (see Körkkö, Kyrö-Ämmälä and Turunen 2016). Furthermore, the chapter provides discussions on recreating Indigenous and emancipatory practices in teaching.

In this study, the teacher of the Pilot project designed and implemented a playful pedagogical innovation aimed at indigenising teaching by strengthening children's voices and agency in Sami language distance education. The teacher recorded the pedagogical innovation, which was viewed together with the researchers in a stimulated recall interview shortly after the recording (Malva, Leijen, and Arcidiacono 2021). These data were analysed using an analytical framework developed by Blaisdell et al. (2019) for the study of young children. The teacher's perception of supporting children's voices and agency in the distance education of the Sami language was examined through the eight aspects of this framework. The aspects were adapted to this study and are explained in more detail later.

Our research questions were formulated based on the aforementioned analytical framework (Blaisdell et al. 2019) to describe how the teacher developed her teaching methods during the innovation study and how this process proceeded. Accordingly, the following four research questions were formulated:

1 How does the teacher define and listen to children's voices in Sami language distance education?
2 How does the teacher process and structure distance teaching's development needs into practical actions in her teaching?
3 How does the teacher promote inclusion and empowerment through playful learning?
4 What skills does the teacher need to implement playful learning and to emphasise children's voice and agency?

Based on the records of the Pilot project, distance learning of the Sami languages will continue to expand as new learners participate in Sami language education, generating a constant demand for expansion and support. This expansion is also supported by increasing expectations for the provision of Sami language distance education; however, Sami language distance education is an understudied domain. Therefore, we based our study on the need to increase knowledge about Sami language learning from an innovative point of view, with the aim of identifying the needs of diverse learners. Previous studies have focussed on the opinions of parents, teachers, and principals (Helander, Keskitalo, and Turunen 2022b). To develop and further scaffold teaching, specifically for children, their voices need to be heard as well (Quinn and Owen 2014).

Playful learning as resistance

Adopting a playful approach towards learning and knowledge to recreate creative and collaborative learning can facilitate ontological change within pupils (Rice 2009). Kangas (2010) noted that in terms of tutoring and lesson planning, teachers feel their role when organising playful learning is important but challenging at the same time. Children consider playful learning insightful from the point of view of co-creation: turning fact into fiction can be a rewarding way to learn, practise group work, and use one's imagination for a common goal (Kangas 2010). A study by Kangas et al. (2017) indicated that teachers' pedagogical and emotional engagement in playful learning partially contributes to pupils' satisfaction.

In general terms, many Indigenous communities portray traditional forms of playing as challenging, resisting, and even displacing colonial agendas (Downey and Neylan 2015). According to Balto (1997), the play consists of a significant socialisation process in which parents, guardians, and adults also play significant roles. Many Indigenous people have appreciated the role of play in child-rearing. Traditional Sami child-rearing, in this sense, is based on the reality that playing has a central role in child-rearing, and therefore it is given a time and place to happen. Play offers possibilities for teaching children on the basis of their initial understanding and gives them a place to build competency for their later lives.

Traditional play also includes possibilities for transforming pedagogies. By observing Sami children play, Balto (1997) found that concerning traditional games, children imagined themselves doing adult activities in traditional livelihoods; these actions help develop values and attitudes as well as prepare them to engage in their own livelihoods later. Parents and adults took part in different forms of play. For example, parents made for their children traditional toys that imitated—in smaller forms—real objects belonging to Sami culture. When children are encouraged to play, they are also driven to develop their cultural and intercultural values (Balto 1990, 1997, 2005).

The aim of Indigenous education is to promote the construction of new paths for social justice practices, resulting in a more culturally diverse educational curriculum (Pereira, Menezes, and Venâncio 2021). In this sense, as a curriculum goal, playing can be understood as part of developing a wide range of skills among pupils (Opetushallitus 2016). Play can also be understood as beneficial due to its potential and capability for creating a social and cultural environment that supports curriculum goals (Santamaria 2009). Playing and playful learning can be referred to as instruments for developing a zest for learning and supporting creative and insightful thinking.

According to Robinson (2019), the development of playful learning activities enriches relationships and enhances children's development. Indigenous language revitalisation efforts benefit from traditional playful artistry, even in an applied form, despite the fact that this process involves obstacles and pouring out hope, like sincere expressions of regret and peace, sadness and optimism (Sadeghi-Yekta 2020). At the same time, ideas regarding how to promote learning also vary across the educational field (Lillard 2013).

Digital solutions indeed bring about and enlarge playful arenas and possibilities for learning (Laiti and Frangou 2019; Plass et al. 2014; Price et al. 2003; Spikol and Milrad 2008). The growing acceptance of digital games has resulted in an interest in developing methods to take advantage of their potential for educational purposes (Plass et al. 2014). We tie play to the indigenising aim to generate self-determination in education (Keskitalo and Olsen 2021). Decolonisation, according to Guerzoni (2020), requires removing colonial influences from an educational system, whereas indigenisation entails incorporating Indigenous aspects into the educational system. Based on previous research, it is not a new practice to conduct playful learning in digital forms and distance education settings. In addition, we know that Sami educators have been very skilled in developing distance education to meet practical demands and challenges (Rasmus and Pautamo 2012).

Pedagogical innovation study and the analytical framework

The method used in this study is intertwined with the ideas of pedagogical innovation in which teachers are regarded as experts in their own pedagogical practices and where professional development occurs simultaneously with increasing pedagogical agency (Berliner 2001). In the ADVOST research project, the collaboration between teachers and researchers devised a new, creative, and innovative way to teach the Sami languages. This collaboration not only enabled the sharing of ideas but also the developing and testing of teaching methods based on Sami culture, which can be adjudged as innovations in a distance-learning environment.

This study involved planning an innovative pedagogical activity by a group of four Sami language distance teachers, followed by one teacher's application of the same. This activity was observed and reflected on by the researchers and was further evaluated and developed together with both the researchers and the teacher. The aim of this innovation was to develop pedagogical practices to strengthen the voices and agency of young children. To achieve this goal, two workshops held in June and August 2020 were attended by the four teachers. During these workshops, the four teachers planned pedagogical innovations to test how storytelling, playful learning and land-based pedagogies might work in an online environment (see Helander et al. 2022b). The action phase of the innovations was implemented during the autumn semester of 2020, while the second phase was conducted during the spring semester of 2021. As mentioned in the introduction, this chapter describes one case and provides insights into one teacher's innovation in playful learning conducted in an online environment.

The data were drawn from one teacher's logbook, written during the innovation, and two audio-recorded stimulated recall interviews conducted after the first and second phases of the innovation in November 2020 and February 2021, respectively. Stimulated recall is a method in which the teacher first records a lesson, after which the recording is watched together with researchers (Schepens et al. 2007). In this study, while watching the recording, the researchers used a semi-structured interview to preserve the teacher's recollection of the ideas and feelings she experienced and the pedagogical choices she made during the lesson.

The data were gathered as a case study, an approach that was chosen due to its usefulness in gaining a deep understanding of real-life phenomena. A case study is needed to identify, explore, and analyse the challenges in the field of education. Thus, in our research, it will not provide a basis for generalisation, validity, reliability, or longevity, which are the challenges (AlBalushi 2019), but will act as an example of how to solve some of the educational and pedagogical challenges in teaching.

The data were analysed by applying Blaisdell et al.'s (2019) analytical framework developed in *Look Who's Talking: Eliciting the Voice of Children from Birth to Seven*—a research project on educational settings where there is diversity in the context of young children's learning. Because the framework was developed for the research of young children's voices in fragile educational settings, it also worked well with the research context of Sami education for uncovering colonial practices and working toward indigenisation while allowing for the identification of complexities regarding learning contexts in young children's educational arrangements. The framework consists of eight aspects—define, listen, process, structure, include, empower, approach, and purpose—in the context of voices of young children. The framework was applied to both the teacher's interview and its analysis to highlight her perceptions of Sami language distance education.

Blaisdell et al.'s (2019) analytical framework was applied as follows:

- *Define* refers to how the teacher defines children's voices and agency in distance teaching.
- *Listen* refers to how the teacher maintains the children's voices and agency in her teaching.
- *Process* refers to developmental needs and issues observed by the teacher that interfere with pedagogical practices.
- *Structure* explains how the playful innovation was practically conducted.

- *Include* refers to how children's voices and agency are promoted through playful learning.
- *Empower* refers to how children are connected with language and culture through playful learning.
- *Approach* describes the skills needed by the teacher to promote children's voices and agency in playful learning.
- *Purpose* refers to the purpose and meaning of the pedagogical innovation for the teacher.

Research ethics are always critical, especially when including Indigenous people as participants (Drugge 2016). It has also been highlighted that Indigenous involvement should be meaningful and should serve the Indigenous community (Juutilainen and Heikkilä 2016)—in this case, the teachers and the children they worked with. The ethics of the current study were approved by an ethical review board at the University of Lapland, which places special attention on data gathering, storing, and handling. The ethics reflect the rules of the Finnish National Board on Research Integrity TENK as well as the regulations followed in the European Union (EU, 679/2016), promulgated by the European Parliament and European Council, also known as the General Data Protection Regulation (GDPR).

This study was conducted through collaboration between researchers and the teacher. Therefore, the teacher who designed and implemented this pedagogical innovation is one of the authors of this chapter. The research group designed the study jointly, collected and analysed the data, and wrote the chapter as part of the ADVOST research project, while the second author designed and implemented the pedagogical experiment in her teaching.

Aspects of playful learning

Define and Listen

The aspect of *define* explains how teachers define children's voices and agency in distance teaching. The teacher reported that she had realised children's motivation should be promoted, as children's learning happens under demanding circumstances. As the supplementary Sami language lessons were already an extracurricular activity for the children, she thought it was important to try motivating them to learn a language that does not have a high status in the surroundings outside their homes and the Sami language online teaching classes. In practice, the children attending Sami language distance education have families living in a diasporic community, where the kin and extended family roots lie often in Northern Finland in the Sami homeland.

The aspect of *listen* explains how the teacher maintains the children's voices and agency in her teaching. The teacher had already devised methods to motivate the children before the innovation study took place. She regularly asked them what they would like to do in the lessons and gave them alternatives. She informed the children that they were able to choose what they would be doing together in the lessons, within certain boundaries. Furthermore, by allowing them to take turns, she tried to account for the wishes of each child. She also chose playful learning to serve the children's voices and agency effectively in a demanding environment (see also Blaisdell et al. 2019).

Process and Structure

The aspect of *process* describes the developmental needs of teaching from the teacher's point of view. The teacher had worked for many years as a Sami language teacher in primary schools in the Sami homeland; however, during the study, she was working as a distance teacher for the first time. Her pupils lived outside the Sami homeland, and their language skills were diverse. At first, the teacher found it challenging to teach children with more passive language skills. Her pupils were young, 6–7 years old, and in addition to language skills, their studying skills were also diverse. The teacher observed that it was quite difficult to organise distance teaching for children of that age group; therefore, it would be meaningful to address their age-specific needs (see Boyle et al. 2010). The teacher made the innovative choice to work with playful learning based on her ideas and the specific challenges she was facing.

The teacher's first interview revealed that plays, songs, and games that provide variations in lessons were already part of her teaching process. When planning the playful learning innovation, the teacher wanted to test whether the whole lesson could be built on play; however, she wrote in her logbook that it was difficult to get started on the innovation planning as it was tough to come up with play that would be seamlessly compatible with distance education.

During the planning phase, the teacher considered several different options for playing and tested how her ideas would work with the children. At first, the teacher thought adjectives could be taught using the "draw and guess" game; however, the idea failed because the Microsoft Teams white board proved difficult to use by children. Eventually, the teacher proposed the idea of toyshop play. Subsequently, when she asked if the children would like to play toyshop, they were very excited. She came up with the idea of using a document camera to implement toyshop play remotely, so the children would actually see the toys in her hands, not just their pictures.

The aspect of *structure* explains how the content of the innovation was practically conducted. The teacher had created a structure for the lessons in which she first began by asking how each child was doing, followed by singing a Sami song together. The purpose of these activities was to lead children to converse in the Sami language. Consequently, when the teacher told the children they would be playing toyshop, they became enthusiastic about sharing their own experiences with play. Although the children continued to speak Finnish at this point, the experience of play prompted them to speak in Sami. The goal of toyshop play was to awaken children's Sami conversation skills as their language skills were still quite passive, and at the same time to practise some vocabulary, especially numbers.

The teacher began play by presenting items for sale through the document camera. The teacher named the items in Sami while the children suggested their prices. Finally, the teacher compiled all the objects and prices for the children to see on the document camera. The children were shopping one by one while the teacher collected the goods they chose into small shopping carts so that they could see the action happening through the document camera. The teacher spoke in Sami throughout play, but the children also had the opportunity to speak in Finnish; however, gradually, they began to repeat phrases in Sami and also received help from each other in the process (Figure 3.1).

At the end of the lesson, the teacher offered some movement exercises to the children, where they repeated numbers in a fortune wheel. The teacher and the children had collected some movements, such as X-jump or flossing, and compiled them into a digital fortune wheel, which randomly picked up one of the movements. Next, the children decided the number of times the movement should occur, after which the teacher and the children together completed the exercise by counting the movements in Sami. This activity had become a routine for every lesson, and whenever the teacher forgot the exercise, the children would ask for it.

Figure 3.1 Sunna Kitti, the Pilot project on distance education in Sámi languages.

The teacher worked on the innovation based on her own evaluation and feedback from the researchers and reimplemented it in February 2021. She decided to add some reading and writing skills exercises at the beginning of toyshop play, as the children's studying skills had already evolved at that point of the school year. She reported:

> This time, I changed this a bit. I made a Power Point presentation with a picture of that toy first; we talked about this toy and said it in Sami. Then, I wrote it in Sami so that they could see the word written. Then, we read it aloud together. So, we were able to include reading as well.

As a result, on seeing the toy, the children also registered how the word for it was written and heard how it was pronounced in addition to reading the word themselves. Therefore, this activity connected many types of learning. Next, the teacher showed the toys one by one through the document camera; the children could see the teacher moving the real toys in her hands, and as she asked the children what it was called, they remembered and repeated the concepts in Sami (see Davis and Kim 2001). Thus, it can be established that playful learning provides opportunities to support and transform teaching in a meaningful way (Kervin 2016).

Inclusion and Empowerment

The aspect of *inclusion* explains how the teacher promoted each child's voice and agency through playful learning. The teacher had spent a lot of time grouping the children. There was a good atmosphere in the group; everyone felt included, and the children encouraged and helped each other to learn Sami. In the first interview, while watching the recording of toyshop play, the teacher recognised that the children were so excited about play that they often talked over each other. She reflected that she could have turned off the microphones of the other children when one was shopping so that each child would have the opportunity to shop in peace.

The teacher highlighted that the children were at the core of her teaching: "I always think of what is best for them and want their motivation and opinions to be at the core" (interview). She found it crucial to empower children and to maintain their status as committed learners, keeping in mind that Sami language teaching is an extracurricular activity. The teacher also found it important to deliver the lessons at a level that the children would enjoy and would feel secure in themselves. According to the teacher, the balance between studying an endangered language and having fun at the same time is important.

As mentioned, play, games, and songs were already part of the teacher's routine in the distance Sami lessons. The children had already played a few available online games in the Sami language. In addition, the teacher herself made vocabulary-learning games in a Wordwall app. The app has several templates, such as memory games, crosswords, and balloon pops, to which the teacher adds words. Such applications help teachers create the appropriate material for distance learning quickly because Sami language teaching lacks such materials due to the lack of economic and human resources (Korpela 2020).

During the first phase of the innovation study, the children wanted to continue the shop play in the next lesson, and it was agreed that they could bring their own toys for sale. The next lesson was a bit different; this time, the children acted as both sellers and buyers, while the teacher stayed in the background, guiding the play when needed. Introducing their own toys to classmates gave the children the opportunity to bring out their own voices and identities during the lesson. According to the teacher, the children's enthusiasm to continue play for the next lesson was the best feedback she received from them. By providing them with the opportunity to continue and develop the play, the teacher included the voices of the children in the planning of the next lesson.

The aspect of *empowerment* describes how the teacher connects children to the language and culture through playful learning. The Sami song *Mun biepmu dutnje attán* that was sung at every lesson talks about the kinds of animals found on Sami land and what they eat. Such kinds of playful learning connect the children to Sami animals, even when they currently live in suburban regions outside the core Sami areas. It is still important to introduce them to the names of animals that are significant to Sami culture, as such knowledge serves as a starting point for discussions on cultural issues.

As the children learnt animal names and what they ate, it became possible to discuss other features of the animals and Sami nature. For example, adjectives can be learnt by discussing whether something is soft, hard, round, square, small, or big, while a range of colours can similarly be taught. The children can also find something similar from their surroundings and share it with others by showing it on the webcam (teacher's logbook). These activities motivate children and imbue learning with attention and interest.

Playful learning in the Indigenous context also seems to influence homes. According to the teacher's logbook, one of the children had decided to buy a CD which had the song they were singing in the lessons so that they could practise it at home as well. If the cultural factors that children encounter at school are meaningful, the effect is felt in their homes as well (Hermes and King 2013). In a previous study of this research group, parents indicated that when children learn playfully during lessons, they learn the Sami language well, particularly if their language competency seems to need some improvement (Helander et al. 2022a).

For toyshop play, the teacher tried to find toys that children would relate to while studying the Sami language. Eventually, she found one toy that resembled the character *Uijui* from the Sami Children's Programme. Although toyshop play may not seem oriented to Sami culture at first, trading has been an important part of Sami culture, and children often accompany adults to markets to learn trade (Harlin 2007). In the interview after the second phase, the teacher expressed that the play could be further developed by using Sami toys and objects for sale.

Approach and Purpose

The aspect of a*pproach* describes the skills required by the teacher to promote children's voices and agency in playful learning. The teacher indicated that without project-based support, it would have taken more time to develop and rethink the teaching strategies. Before this joint effort, she had no time or capacity to carry this out alone at the level she wanted: to be consistent, well-prepared, aware, and confident in the meaningfulness of the planned activities. The teacher seemed to need confirmation that her teaching methods were helping the children, so mutual collegial support seemed necessary. In this sense, this kind of a pedagogical innovation study allowed the teacher's expertise to grow.

The teachers of the Pilot project received training on utilising digital applications in their innovations. After the second phase of the innovation, the teacher highlighted in her interview that although she acquired some new teaching ideas from training, she did not feel comfortable using these applications:

> I did not eventually use those programmes or applications. I tried them after a course that was given to us, but it was outside my comfort zone, so despite the good aim, I eventually used only PowerPoint.

In practice, this means that while teachers receive support and training to enhance their digital competencies, it could be somewhat challenging to start using different applications. It seems that for the teacher to use more complex and advanced solutions, a colleague-teacher partnership may be required so that the one using different solutions can be a part of the teaching as well and subsequently create something collaboratively (e.g. Tanhua-Piiroinen et al. 2016). This could be focused on in future research involving these kinds of experiments.

The teacher acknowledged in her logbook that changing and developing teaching techniques requires concentration, time, and energy. In addition, competency must be developed securely. Therefore, as this study demonstrates, teachers need knowledge about children's learning conditions and bases as well as the requirements for working in diverse and complex teaching situations. Based on these complicated and demanding settings, teachers also need comprehensive knowledge of digital solutions and their use (Tanhua-Piiroinen et al. 2016).

The aspect of *purpose* explains the purpose and meaning of the pedagogical innovation to the teacher. The teacher felt that the more she played with young children, the easier it was to transition them to the world of older children, as younger ones are more receptive to new experiments. She felt that the innovation study enabled her to find the courage to try new things in her teaching. According to the teacher's logbook, a great deal of thinking went into her actions during the interviews and the development processes of the pedagogical innovation. Such reflective development work creates the possibility to work in alternative scenarios, provides the opportunity to become more knowledgeable, and helps to exercise broad critical judgments (Day 2012).

Conclusion

This chapter dealt with playful learning in the context of Sami language learning and distance education. Although playing is a form of traditional child-rearing practice in Sami culture (Balto 1990), it poses challenges when being incorporated into learning. The teacher in this study realised that in the course of her daily teaching, she faced challenges in helping her pupils make progress. Accordingly, she needed to consider playful learning—not only because it is often the best choice when working with children but also because it is an integral part of Sami culture.

In the current pedagogical innovation developed by the teacher, the mode of playfulness was chosen from modern life depending on something which is familiar to children from their lives in suburban areas: shopping. While playing at shopping, the children used familiar concepts and words, which was a mindful approach to learning. In the future, the question remains as to how this could be integrated in the teaching of the Sami language, including the aspects that arise from the culture itself. The teacher in this study already had solutions to address such a question, such as by including lessons on animals that are somewhat familiar to the children and playing within a thematic field.

This retrospective analysis highlighted ways in which innovation in education can be modified to fit the needs of language communities, inform language revitalisation efforts, and assist with the evolution of community-based research designs. Broadly, the praxis described in this article draws on community collaboration, knowledge production, and the evolution of a design for Indigenous language revitalisation (see also Hermes et al. 2012).

All things considered, this chapter describes the indigenisation of education by building bridges between traditional Sami child-rearing practices and institutionalised learning engaging decolonisation of education. This means that teaching needs to be reorganised in a way by being more inclusive and representative of Indigenous peoples, perspectives, and place (Guerzoni 2020). It is our hope that this research will enable teachers and children to engage with the complex Sami language distance education environment more effectively despite its many shortcomings. As presented in this chapter, the pedagogical innovation serves as an example for teachers to create learning environments for children that are motivating, based on their educational starting points and their desires, and provide several practical ways to work with the endangered Sami language in a setting where diverse learners need inclusive solutions. We encourage other researchers and educational practitioners to broadly cooperate and collaborate to solve such educational challenges.

Notes

1 The research presented in this article was enabled by two projects: 1) the Pilot project on distance education in the Sámi languages funded by the Finnish Ministry of Education; and 2) Culture and Socially Innovative Interventions to Foster and to Advance Young Children's Inclusion and Agency in Society through Voice and Story (ADVOST) funded by the Academy of Finland. This work was supported by the Academy of Finland under Grant 334791.
2 Basic Education Act – Finland: Perusopetuslaki 21.8.1998/628. https://www.finlex.fi/fi/laki/ajantasa/1998/19980628
3 The Pilot Project on distance education in Sámi languages: https://www.saamenetaopetus.com/hanke
4 Socially Innovative Interventions to Foster and to Advance Young Chidren's Inclusion and Agency in Society through Voice and Story (ADVOST): https://www.teachered-network.com/projects/advost/

References

AlBalushi, Zaaima Talib. 2019. "Challenges of a Case Study." In *Case Study Methodology in Higher Education*, edited by Annette Baron and Kelly McNeal, 323–343. Hershey, PA: IGI Global.

Arola, Laura. 2020. *Selvitys saamenkielisen opetus-ja varhaiskasvatushenkilöstön saatavuudesta ja koulutuspoluista*. Helsinki: Opetus- ja kulttuuriministeriö. https://julkaisut.valtioneuvosto.fi/handle/10024/162556

Balto, Asta. 1990. "Samisk barnelek i 1980-årene. Eksempler fra en barnehage [Children's play in a Sami day-care center]." *Tradisjon* 10: 89–96.

Balto, Asta. 1997. *Sámi mánáid bajásgeassin* [Sami childrearing practices in change]. Oslo: Ad Notam Gyldendal.

Balto, Asta. 2005. "Traditional Sámi Child Rearing in Transition: Shaping a New Pedagogical Platform." *AlterNative: An International Journal of Indigenous Peoples* 1 (1): 85–105.

Berliner, David C. 2001. "Learning About and Learning From Expert Teachers." *International Journal of Educational Research* 35 (5): 463–482.

Blaisdell, Caralyn, Lorna Arnott, Kate Wall, and Carol Robinson. 2019. "Look Who's Talking: Using Creative, Playful Arts-based Methods in Research with Young Children." *Journal of Early Childhood Research* 17 (1): 14–31.

Boyle, Frank, Jinhee Kwon, Catherine Ross, and Ormond Simpson. 2010. "Student–student Mentoring for Retention and Engagement in Distance Education." *Open Learning: The Journal of Open, Distance and e-Learning* 25 (2): 115–130.

Brzozowski, Alexandra. 2021. Nordic Countries Set Up Sámi Reconciliation Commissions to Investigate Indigenous Injustices. Euractiv. Where is discrimination in Europe? Special report: 12–13. https://en.euractiv.eu/wp-content/uploads/sites/2/special-report/Where-is-discrimination-in-Europe_-Special-Report-1.pdf

Davis, Chris, and Jeesun Kim. 2001. "Repeating and Remembering Foreign Language Words: Implications for Language Teaching Systems." *Artificial Intelligence Review* 16 (1): 37–47.

Day, Suzanne. 2012. "A Reflexive Lens: Exploring Dilemmas of Qualitative Methodology Through the Concept of Reflexivity." *Qualitative Sociology Review* 8 (1): 60–85.

Downey, Allan, and Susan Neylan. 2015. "Raven Plays Ball: Situating 'Indian Sports Days' Within Indigenous and Colonial Spaces in Twentieth-century Coastal British Columbia." *Canadian Journal of History* 50 (3): 442–468.

Drugge, Anna Lill (ed.). 2016. *Ethics in Indigenous Research: Past Experiences - Future Challenges*. Vaartoe-Centre for Sami Research. Umeå: Umeå University.

Geia, Lynore, and Melissa Sweet. 2013 "#IHMayDay: Showcasing Indigenous Knowledge and Innovation." In *Proceedings of the 13th National Rural Health Conference*. www.ruralhealth.org.au

Guerzoni, Michael A. 2020. *Indigenising the Curriculum: Context Concepts and Case Studies*. University of Tasmania. https://www.utas.edu.au/__data/assets/pdf_file/0019/1452520/Indigenising-the-Curriculum-Context-Concepts-and-Case-Studies.pdf

Harlin, Eeva-Kristiina. 2007. "Suomen puoleisen Tornion Lapin markkinat." In *Peurakuopista kirkkokenttiin: saamelaisalueen 10 000 vuotta arkeologian näkökulmasta*, edited by Eeva-Kristiina Harlin, and Veli-Pekka Lehtola, 154–167. Oulu: University of Oulu.

Helander, Hanna, Pigga Keskitalo, Sirkka Sanila, and Sonja Moshnikoff. 2022a. "Saamelaiskulttuurilähtöinen tarinankerrontainnovaatio vahvistamssa lapsen osallisuutta omaan kulttuuriin saamen kielen etäopetuksessa." *Dutkansearvvi dieđalaš áigečála* 6 (1): 50–71. https://www.dutkansearvi.fi/wp-content/uploads/Volume6Issue1-02-3-Helander-Keskitalo-Sanila-Moshnikoff-Finnish-web.pdf

Helander, Hanna, Pigga Keskitalo, and Tuija Turunen. 2022b, in print. "Saami Language Online Education Outside the Saami Homeland: New Pathways to Social Justice." In *Finland's Famous Education System - Unvarnished Insights into Finnish Schooling*, edited by Martin Thrupp, Piia Seppänen, Jaakko Kauko, and Seppo Kosunen. New York: Springer.

Hermes, Mary, Megan Bang, and Amanda Marin. 2012. "Designing Indigenous Language Revitalization." *Harvard Educational Review* 82 (3): 381–402.

Hermes, Mary, and Kendall A. King. 2013. "Ojibwe Language Revitalization, Multimedia Technology, and Family Language Learning." *Language Learning and Technology* 17 (1): 125–144.

Juutilainen, Sandra, and Lydia Marja Terttu Heikkilä. 2016. "Moving Forward with Sámi Research Ethics: How the Dialogical Process to Policy Development in Canada Supports the Course of Action for the Nordic Countries." In *Ethics in Indigenous Research: Past Experiences–Future Challenges*, edited by Anna-Lill Drugge, 81–104. Umeå: Vaartoe-Centre for Sami Research.

Kangas, Marjaana. 2010. "Creative and Playful Learning: Learning Through Game Co-creation and Games in a Playful Learning Environment." *Thinking Skills and Creativity* 5 (1): 1–15.

Kangas, Marjaana, Pirkko Siklander, Justus Randolph, and Heli Ruokamo. 2017. "Teachers' Engagement and Students' Satisfaction with a Playful Learning Environment." *Teaching and Teacher Education* 63: 274–284.

Kervin, Lisa K. 2016. "Powerful and Playful Literacy Learning with Digital Technologies." *Faculty of Social Sciences - Papers*. 2879. University of Wollongong. https://ro.uow.edu.au/sspapers/2879

Keskitalo, Pigga, Kaarina Määttä, and Satu Uusiautti. 2012. "Sámi Education in Finland." *Early Child Development and Care* 182 (3–4): 329–343.

Keskitalo, Pigga, and Torjer Olsen. 2021. "Indigenizing Education: Historical Perspectives and Present Challenges in Sámi Education". In *Arctic Yearbook 2021 – Defining and Mapping the Arctic*. edited by Lassi Heininen, Heather Exner-Pirot, and Justin Barnes, 469–478. Iceland: Arctic Portal. https://issuu.com/arcticportal/docs/ay2021

Körkkö, Minna, Outi Kyrö-Ämmälä, and Tuija Turunen. 2016. "Professional Development Through Reflection in Teacher Education." *Teaching and Teacher Education* 55: 198–206.

Korpela, Helena. 2020. *Saamenkielisten oppimateriaalien monet kasvot: Selvitys saamenkielisen oppimateriaalin tilanteesta ja tulevista tarpeista.* Helsinki: Opetus- ja kulttuuriministeriö.

Kuokkanen, Rauna. 2000. "Towards an 'Indigenous Paradigm' from a Sami Perspective." *The Canadian Journal of Native Studies* 20 (2): 411–436.

Laiti, Outi K., and Satu-Maarit Frangou. 2019. "Social Aspects of Learning: Sámi People in the Circumpolar North." *International Journal of Multicultural Education* 21 (1): 5–21.

Lehtola, Riitta, and Pia Ruotsala. 2017. *Saamenkielisten palveluiden nykytilakartoitus Saamelaisten lasten-, nuorten ja perheiden palvelut.* Inari: Saamelaiskäräjät. https://stm.fi/documents/1271139/4067344/SAAMELAPEselvitys110117_final.pdf/f1418169-7e37-4d7a-803d-30d192ffed5f

Lehtola, Veli-Pekka. 2015. "Sámi Histories, Colonialism, and Finland." *Arctic Anthropology* 52 (2): 22–36.

Lillard, Angeline S. 2013. "Playful Learning and Montessori Education." *NAMTA Journal* 38 (2): 137–174.

Malva, Liina, Äli Leijen, and Francesco Arcidiacono, 2021. "Identifying Teachers' General Pedagogical Knowledge: A Video Stimulated Recall Study." *Educational Studies*: 1–26. https://doi.org/10.1080/03055698.2021.1873738

Nyyssönen, Jukka. 2013. "Sami Counter-narratives of Colonial Finland. Articulation, Reception and the Boundaries of the Politically Possible." *Acta Borealia* 30 (1): 101–121.

Opetushallitus. 2016. *Perusopetuksen opetussuunnitelman perusteet 2014.* 4th. ed. Helsinki: Opetushallitus. https://www.oph.fi/sites/default/files/documents/perusopetuksen_opetussuunnitelman_perusteet_2014.pdf

Pasanen, A. (2016). Saamebarometri 2016. *Selvitys saamenkielisistä palveluista saamelaisalueella-Sámi giellabaromehter 2016. Čielggadus sámegielat bálvalusain sámeguovllus.* Helsinki: Oikeusministeriö. https://julkaisut.valtioneuvosto.fi/bitstream/handle/10024/78941/OMSO_39_2016_Saamebaro_120s.pdf?sequence=1

Pereira, Arliene, Stephanie Menezes, and Luciana Venâncio. 2021. "African and Indigenous Games and Activities: A Pilot Study on Their Legitimacy and Complexity in Brazilian Physical Education Teaching." *Sport, Education and Society*, 26 (7): 718–732.

Plass, Jan L., Bruce D. Homer, and Charles K. Kinzer. 2014. "Playful Learning: An Integrated Design Framework." *White Paper* 2.

Price, Sara, Yvonne Rogers, Mike Scaife, Danae Stanton, and H. Neale, 2003. "Using 'tangibles' to Promote Novel Forms of Playful Learning." *Interacting with Computers* 15 (2): 169–185.

Quinn, Sarah, and Susanne Owen. 2014. "Freedom to Grow: Children's Perspectives of Student Voice." *Childhood Education* 90 (3): 192–201.

Rasmus, Eeva-Liisa, and Ellen Pautamo. 2012. "Oikeus omaan kieleen ja kulttuuriin." In *Apuja aktiivisuuteen, välineitä verkostoihin. Apuja aktiivisuuteen, välineitä verkostoihin*, edited by M. Sihvonen and K. Saloniemi, 35–39. Hämeen ammattikorkeakoulu.

Rice, Louis. 2009. "Playful Learning." *Journal for Education in the Built Environment* 4 (2): 94–108. https://doi.org/10.11120/jebe.2009.04020094

Robinson, Jenny Perlman. 2019. *Philadelphia Playful Learning Landscapes: Scaling Strategies for a Playful Learning Movement.* Washington, DC: Center for Universal

Education at The Brookings Institution. https://www.brookings.edu/research/philadelphia-playful-learning-landscapes/

Sadeghi-Yekta, Kirsten. 2020. "Drama as Methodology for Coast Salish Language Revitalization." *Canadian Theatre Review* 181: 41–45.

Salminen, Tapani. 2010. "Europe and the Caucasus". In *Atlas of the World's Languages in Danger*, edited by C. Moseley, 32–42. UNESCO. https://unesdoc.unesco.org/ark:/48223/pf0000187026/PDF/187026eng.pdf.multi

Santamaria, Lorri J. 2009. "Culturally Responsive Differentiated Instruction: Narrowing Gaps Between Best Pedagogical Practices Benefiting all Learners." *Teachers College Record* 111 (1): 214–247.

Schepens, Annemie, Antonia Aelterman, and Hilde Van Keer. 2007. "Studying Learning Processes of Student Teachers with Stimulated Recall Interviews Through Changes in Interactive Cognitions." *Teaching and Teacher Education* 23 (4): 457–472.

Spikol, Daniel, and Marcelo Milrad. 2008. "Physical Activities and Playful Learning Using Mobile Game." *Research and Practice in Technology Enhanced Learning* 3 (03): 275–295.

Tanhua-Piiroinen, Erika, Jarmo Viteli, Antti Syvänen, Jaakko Vuorio, Kari A. Hintikka, and Heikki Sairanen. 2016. "Perusopetuksen oppimisympäristöjen digitalisaation nykytilanne ja opettajien valmiudet hyödyntää digitaalisia oppimisympäristöjä." *Valtioneuvoston selvitys- ja tutkimustoiminta*. Helsinki: Valtioneuvosto. https://julkaisut.valtioneuvosto.fi/bitstream/handle/10024/79573/perusopetuksen%20oppimisymp%C3%A4rist%C3%B6jen%20digitalisaation%20nykytilanne.pdf?sequence=1

4 The virtue of extraction and decolonial recollection in Gállok, Sápmi

Georgia de Leeuw

Introduction

Sweden enjoys a favourable international reputation and discursively enacts a long list of national exceptionalisms pertaining to high standards in human rights, social welfare, equality, and environmentalism. This chapter discusses Sweden's mining identity as an additional avenue of exceptionalism, while highlighting concerns of indigenous injustice and environmental degradation voiced against its expansive mining industry. Pro-mining advocates on state, municipal, and industry levels narrate Swedish resource richness and mining as a source of pride. It has helped overcome the Swedish famine, finance its welfare state, and it manifests a promise of the modern way of life. Swedish extraction is here envisioned exceptional, virtuous, as coinciding harmoniously with quests for sustainability and the cultural survival of the indigenous Sami community. This imagined virtue is especially apparent in the Swedish iron and steel industry, which has functioned as an important incentive for historic and current advances to and through the North. In the words of former Prime Ministers Fredrik Reinfeldt and Stefan Löfven:

> Our mining industry and our iron ore is for us what oil is for the Norwegians. An amazing prosperity, an opportunity to build investments and development for the future [...][1]
>
> (Reinfeldt, Sveriges Radio, October 26, 2012)

> Steel has built Sweden, and steel has built our welfare.[2]
>
> (Löfven, Government Offices of Sweden, August 31, 2020)

I read this envisioned extractive exceptionalism as a linear imaginary of past and current activities in Sápmi which require disregard for stories of the margins. Post- and decolonial scholars have provided insight into silences of the "embodied history" of the colonised mind (Hernández Castillo 2020; Thiong'o 1987), or that which has come to be sheltered in the "heads and hearts of people", in rules, policies, and commonsensical knowledge (Wekker 2016, 19). Colonial and postcolonial states often display difficulties to

DOI: 10.4324/9781003293323-5

accommodate their ties to colonial expansion in their national self-imaginaries (Donadey 1999). Carl-Gösta Ojala and Jonas Nordin (2019) also recognise these tendencies of memory loss in the Swedish self, whereby Swedish colonialism is often imagined exceptional, "somehow 'kinder' and less 'colonial' than that of other empires" (2019, 102). Instead, scholars and activists have argued that Swedish expansion in Sápmi is far from exceptional as it resembles a rather routine form of colonial policy (Lindmark 2013; Össbo and Lantto 2011).

Inspired by such accounts, this chapter posits that Sweden's sense of exceptionalism in mining is rooted and embedded in a colonial/extractive system of knowledge which fails to mention non-extractive, non-capitalist, anti-colonial relationalities to nature and ways of life. Its dominant status helps to justify and naturalise extractive impulses that "serve the purposes and needs of capitalism" (Quijano 2000, 550; see also Lander 2002), and which materialise in transformations of indigenous land into mining sites. In this chapter, I aim to examine Swedish stories of exceptional resource rich bedrocks, non-intrusive, just, and sustainable mining in Swedish administered Sápmi and the colonial/extractive rationale that these narratives help reproduce. The purpose is to contribute to decolonial literature with an analysis of the colonial forgetfulness (Maldonado-Torres 2004) that upholds the coherence of Swedish exceptionalism, and the disruptive power of what I refer to as *decolonial recollection* in challenging the dominant extractive episteme. I show how narratives pushed to the margins can remind the dominant imaginary of its colonial tendencies and environmental degradation and thereby disrupt forgetful storylines. The chapter also contributes to literature on Swedish exceptionalism by discussing the construction of extractivism as virtue, highlighting its ties to colonial innocence. With the above aim in mind, I ask: how is Swedish exceptionalism constructed, upheld, and disrupted in land claims? And, how do instances of decolonial recollection function to disrupt such imagined exceptionalism?

A narrative reading of epistemes

I analyse the reproduction of this colonial/extractive episteme and its contestations that come in the form of non-extractive, anti-capitalist, anti-colonial narrative disruptions (see Galván-Álvarez 2010). I read competing epistemic structures through their narrative iterations by pro- and anti-mining actors, respectively. I understand narratives as stories that we tell to make sense of our surroundings. Thus, I perceive state, municipal, corporate as well as indigenous and local actors as constructing narratives of land and mining that either reinforce or challenge existing knowledge structures (Patterson and Monroe 1998). It should be noted that I do not seek to allocate actors individual narratives, but to identify the dominant idea of extractive exceptionalism as (re)enacted by pro-mining, and challenged by anti-mining, voices. Still, I acknowledge that actors may become complicit (Vuorela 2009) in epistemic violence through narrative assertion of the colonial/extractive episteme. I provide a discursive reading of diverse forms of texts and visuals (Rose 2001), including documents, websites, photographs, art, and videos. While narratives are commonly associated with text or speech, I recognise the discursive potential of images and visuals in telling stories. The visuals used are read as having agency in delineating and framing reality (Butler 2010), yet are also read intertextually and contextually for additional interpretive opportunities. This allows me to read visuals and text as narratives that tell diverse stories of mining.

A locality of extraction

I illustrate extractive virtue and decolonial recollection, respectively, through the locality of Gállok/Kallak[3] in the reindeer herding areas of Sirges and Jåhkågasska Sami villages or Jokkmokk municipal district. Gállok is located on a peninsula in the Lesser Lule River and is identified as an area of national interest for both its mineral deposit and reindeer husbandry. The pro-mining side—consisting of state, government and opposition actors, municipal council representatives, the Swedish mining industry, and British Beowulf Mining—sees Kallak as housing exceptional iron ore deposits, and with it opportunities for growth and regional transformation. Sami, local and environmental opposition see a cultural landscape threatened by colonial theft and environmental degradation. In the case of Gállok, the taken-for-granted idea of the North as a frontier of extraction has been challenged, which is visible in a permit process in which decisions were postponed for years. At the time of writing, the government's decision has finally been announced by mine-enthusiast and Minister of Business Karl-Petter Thorwaldsson, giving the go-ahead for the project on 22nd March 2022. Still, with an environmental assessment pending, the Sami and local resistance does not seem to view the decision as the end of the road. Indicative of Gállok's controversial status is also that the permit came with an unprecedented list of conditions involving, among others, limitations to the land area used, the restoration of land post-project, and the assurance of reindeer transports in case herding routes are made inaccessible. This chapter focuses on the time leading up to the decision.

The chapter is outlined as follows. First, I account for theoretic literature on exceptionalism. I then discuss the dominant colonial/extractive episteme and the epistemic violence that is exerted through its reproduction, where I also introduce the framework that I propose—*colonial forgetfulness/decolonial recollection*. After a brief case description, the analysis is presented in two parts. The first dissects the forgetfulness of state, municipal, mining industry, and corporate accounts of exceptional mining. Here, I show how upholding the coherence of Swedish exceptionalism requires a strategic dismissal of anti-mining and Sami knowledge. In the second part, I analyse instances of decolonial recollection that disrupt the linearity of imagined exceptionalism in the case of Gállok.

Map 4.1 Sápmi: The Sami Homelands, Jordan Engel (The Decolonial Atlas n.d.). Gállok is located in Julevsápmi (Lule Sami).

On Swedish exceptionalism

As mentioned above, Sweden is often narrated as manifesting a "gold standard" (Brysk 2009, 42) against which other states' performance should be measured. The markers of Swedish exceptionalism are manifold and may refer to its acclaimed human rights leadership, high level of welfare and democracy, international solidarity, LGBTQI and gender justice, environmentalism, political neutrality, colonial innocence, and anti-discrimination. Swedish exceptionalism is habitually reinforced in policy, academic, and everyday discourse. Yet, critical scholars challenge this "sanctioned ignorance" (Habel 2012), not least toward racial discrimination and colonialism in Sweden (Hübinette 2013). It has been argued that this exceptional imaginary has allowed little room for self-scrutiny, in that "[t]he Swedish story has skipped over less commendable aspects of its history that are inconsistent with prevalent understandings of the self" (Bergman Rosamond 2020, 70). Others have pointed to Sweden's colonial neutrality as a myth by inferring "colonial complicity" in European imperialism (Keskinen et al. 2009). Sweden and its neighbours have also been discussed as rather successful in preserving an image "untouched by colonial legacies" (Mulinari et al. 2009, 18). This despite Sweden's pre-1945 influential role in race biology, its sterilisation programme, colonial policy, and treatment of the Sami peoples. This construction of the Swedish and Nordic colonial past as more innocent than that of "actual" colonisers, "[…] enabled Scandinavia to emerge in the modern period as untainted by colonialism and thus in a position to claim trustworthiness as mediators and as champions of subaltern and minority rights" (Fur 2013, 26).

In her study of the Swedish colony in St. Barthelemy, Lill-Ann Körber maintains that Sweden has been represented in a way that forges impressions of "Swedish colonialism as harmless, good-natured, or even absurd" (2019, 89), allowing for a narrative of innocence stressing the ineffectiveness and irrelevance of Sweden's colonial activities and reach compared to others, all of which eclipsed colonialism from the Swedish conscience altogether. As such, Swedish mining policies were successfully disassociated from colonialism, making any expansion, investment, or settlement a purely Swedish matter, with Sápmi being constructed as an entity submerged under the realm of the Crown (Fur 2013). Sweden has historically disentangled its presence in Sápmi with "that which is commonly referred to as colonialism" (Swedish Ministry of Justice 1986, 164, own translation). As we shall see, this form of reasoning continues to be prevalent in the naturalised targeting of the North as a Swedish resource pool of opportunity. In fact, Rebecca Lawrence and Mattias Åhrén argue that Sweden's inability to acknowledge its colonial history paves the way for today's "increasing non-recognition of Sami land rights" (2017, 152). The Sami people have "little real influence over if and how mining and exploration takes place on their lands" (2017, 157), which is also reflected in the unwillingness to embrace the right to indigenous self-determination through ILO169.[4] Here, Mörkenstam (2019) notes that the Swedish partial endorsement of the indigenous rights regime and international norms is de-coupled from action, which he conceptualises as "organised hypocrisy," allowing for a symbolic recognition of rights without the need of altering policy. This façade of land rights renders northern Sweden a spacious and unexploited area available for investment and development projects (Lawrence 2014). The historic construction of Sápmi as "Swedish America" and "Land of the Future" (Sörlin 1988) is reproduced in dominant accounts of mining today, reinforcing the colonial/extractive episteme.

The colonial/extractive episteme & its disruptions

After having familiarised ourselves with the notion of exceptionalism and its critiques, I here turn to the dominant episteme and its forgetful tendencies. While conflicts over land often imply physical forms of violence—racialised violence of craniometric measurements, profiling, genocide, forced assimilation, or extractive violence (Lawrence 2014; Sehlin MacNeil 2017)—decolonial interventions have pointed to the silencing of knowing, feeling, and remembering otherwise (Drugge 2016; Sandström 2020). Such contributions show that land conflicts and epistemic violence are intimately connected by emphasising that colonial settlements expand not only through lands but also minds, with the enclosure and seizure of land conditioned by and through epistemic violence. *Epistemic violence* (Spivak 2010) refers to "violence exerted against or through knowledge" (Galván-Álvarez 2010, 12). It naturalises the neglect, delegitimisation, and dismissal of certain understandings for the benefit of others (Brunner 2020). What I refer to as the *colonial/extractive episteme* sustains a utilitarian capitalist-nature relationality to land, the logic of domination, taming, ordering, possessing, and cultivating the unexplored, underdeveloped wild, seizing its full potential and value through the mining of its resources (Escobar 1999; Harvey 1996; Merchant 1980). Its elevated epistemic status is reproduced and re-enacted in taken-for-granted narratives of progress, expansion, and exceptionalism, resulting in a legitimised rejection of the epistemically disobedient. What others have themed an "extractive episteme," or "colonial visual regime" (Gómez-Barris 2017), then, marginalises other ways of knowing and relating to land, legitimising further expansionist, extractivist surges.

This dominant episteme is ordered on the basis of the "coloniality of power" (Quijano 2000), which continues to control and define appropriate ways of relating to land, appropriate places for extraction, the expansion of markets through a commodification of nature, made possible with a side-lining of the marginalised, the "wretched of the earth" (Fanon 2001). Coloniality, or what has come to be understood as "the darker side of Western modernity" (Mignolo 2011), is conveniently neglected in the telling of the Swedish virtue of mining. Walter Mignolo (2011) suggests here that the story that is commonly told of modernity adopts a purely "Western" perspective as it excludes its ties to colonial expansionism, subordination, racialised impositions of inferiority, and control over labour and resources. Neglecting the darker side of the coin of coloniality/modernity warrants an imaginary of a "glorious march of modernity" (Tlostanova & Mignolo 2012, 37). It forges an absence of the experiences of "the Other" from the exceptionality of Western progress (Quijano 2000), and positions other ways of knowing opposite of the modern/rational—as irrational. Boaventura De Sousa Santos argues that Western thinking takes an abyssal character that divides modernity/other and renders all that lies beyond that line invisible and indeed "utterly irrelevant" (2014, 71) for the hegemonic gaze. The notion of the "forgetfulness of coloniality" (Maldonado-Torres 2004) makes the absences through which epistemic violence is exerted especially visible, pointing to the dominant episteme's power to silence perspectives of the non-aligned.

Leaning on the debates above and Nelson Maldonado-Torres especially, I argue that through its forgetful tendencies the colonial/extractive episteme continues to sideline "irrational" systems of knowledge. My position here is that safeguarding Sweden's sense of exceptionalism—pertaining to its extractive practices and colonial innocence—necessitates a strategic rejection of anti-colonial, anti-extractive, anti-capitalist experiences and knowledges. Allowing such contradictive accounts to be read and heard, compromises the coherence of the colonial/extractive episteme on which the idea of Swedish exceptionalism is built, and thereby the practices of mining expansion that such imaginary permits. As we shall see, Sami and anti-mining accounts of colonial theft, historical trauma, and the intrusiveness of extractive advances are silenced and dismissed as irrational in the quest for progress, economic growth, and the Swedish virtue of mining. I understand this discursive neglect (or colonial forgetfulness) as a form of epistemic violence that effectively privileges colonial and extractive over anti-mining systems of knowledge. It imposes inferiority on perspectives that threaten to disrupt the sense of exceptionalism, and with it, the extractive capitalist system on the basis of which a modern and sustainable future is envisioned.

Still, resistance in the case of Gállok powerfully disrupts efforts to forget, in that its provocations actively remind and "recollect" that which has been cast aside by the forgetfulness of imagined exceptionalism. I read this subversive recollection as an attempt to repair the "colonial wound" (Tlostanova and Mignolo 2012, 36) that remains open and is ripped anew in instances of dismissal. Existing debates have in decolonial disruptions of coloniality stressed a need to learn to "unlearn" the given order and its colonial legacies (Tlostanova and Mignolo 2012), or elsewhere to "delink" from assumptions of Eurocentrism (Mignolo 2007). Others have spoken of a "decolonisation of the mind" by resisting that which Ngũgĩ wa Thiong'o (1987) has referred to as "colonial alienation," a condition in which the language of the coloniser is accepted as preferable, superior. Others again, stress the importance to "disconnect" from embodied histories and reconnect to others by "putting heart into things" (Hernández Castillo 2020 in dialogue with Xuno López Intzin), thereby detaching from Western-imposed ideas of "rationality" in knowledge production, a logic through which non-Western perspectives continue to be othered and dismissed.

I contribute to such debates with the concept of *decolonial recollection*, which refers to a tapping on the shoulder of stubborn forgetfulness, a disruption of that which safeguards imagined exceptionalism in an attempt to remind the exceptional extractive self that there, beyond its gaze and linear storytelling, lie perspectives and memories of the neglected. Thus, to "recollect" implies posing a subversive reminder that indeed may cause discomfort within the as remarkable imagined self. But to "*re*-collect" also implies a process of recovering, gathering that which has been scattered, lost, discarded, cast aside, a process of reassembling that which has been broken apart, disintegrated. I see in the concept of decolonial recollection a promise of incorporating parts of history which have been decidedly left out. In this *decolonial recollection*—in the reminder of the subversive reassembling of the unaccounted—lies power to incite self-scrutiny of imagined exceptionalism. Decolonial recollection is therefore not following other, indeed powerful, claims for pluriversality and ontological coexistence (De la Cadena and Blaser 2018; Escobar 2018), but instead is a disruptive force, creating discomfort in claims for exceptionalism by requesting the dominant gaze to shift and "look around rather than ahead" (Tsing 2015, 22).

Swedish mining & Gállok/Kallak

Gállok/Kallak is located near the city of Jokkmokk in Norrbotten, the reindeer herding lands of the Sami villages of Sirges and Jåhkågasska tjiellde, and in close proximity to the UNESCO world heritage site of Laponia. In the 1970s, the state agency for bedrock, soil, and groundwater (SGU[5]) identified Kallak as housing the largest unearthed iron ore deposit in Sweden. British mineral extraction company Beowulf Mining has since 2006 been granted several exploration concessions to evaluate the deposit, and has from 2013 until March 2022 waited for a decision on an exploitation permit (Beowulf Mining 2017), involving numerous rounds of statements amongst others by the county administrative board, the Mining Inspectorate of Sweden (Bergsstaten), and UNESCO, accounting for diverse views on the environmental and societal risks versus benefits involved. Beowulf Mining presents Kallak as its flagship project due to "superb drilling results" (Sinclair-Poulton 2012). The case of Gállok/Kallak has risen in controversy after former executive chairman of Beowulf Clive Sinclair-Poulton during a mining convention wondered "What local people?" when asked about potential impacts (Tuorda 2014), causing outrage among the local and Sami community.

Sweden holds a reputation as an easily accessible market of land for mining, and the government vocally advocates for an expansion of the mining industry. Sweden has been listed as one of the most attractive destinations for mining investments, evaluated on the basis of geological and economic extractive potential and attractive mining policies (Fraser Institute, February 25, 2020). Still, recent delays in the permit process (notably in Kallak) have led mining advocates to fear for Sweden's favourable mining reputation (Hjälmered 2020). Meanwhile, government representatives reference the importance of attention to human rights and environmental concerns as a cause of delay (Baylan 2020). The Committee on the Constitution[6] has in late 2020 examined the government's "inefficiency" in the Kallak permit process and concluded that the processing time of seven years, three of which displayed no visible action, was "unacceptable" (KU 2020). Prior to the verdict, the government requested a statement from the Swedish National Commission for UNESCO on the risks associated with mining in Kallak. It responded on June 8, 2021, that an initiation of the mine may constitute a threat to Laponia and its cultural and environmental heritage. Still, pro-mining critics argue that there is no legal need to involve UNESCO since Kallak is situated outside Laponia. In response, Minister for Business Ibrahim Baylan stated that Sweden cannot risk renewed critique from the UN Committee on the Elimination of Racial Discrimination (CERD) for its insufficient dialogue with the Sami people (Baylan 2020). After a shake-up of the government with the Green Party moving into opposition in November 2021, newly assigned Social Democrat Minister of Business Karl-Petter Thorwaldsson has voiced the need to accelerate the permit process, promising a swift decision while reiterating that, "we love mines" (SVT, January 23, 2022). This has led to renewed widespread critique and media attention. On March 22, 2022, he announced the government decision to grant Beowulf mining its long-awaited exploitation permit. The analysis that follows focuses on the narrative iterations leading up to the decision.

Colonial/extractive forgetfulness: the virtue of extraction

Having discussed the chapter's theoretical underpinnings and case, I now analyse state, municipal, industry, and corporate pro-mining narratives that enact the imaginary of Swedish mining as exceptional in attempts to legitimise further expansion through Sápmi. I identify these narratives as reproductions of the colonial/extractive episteme. The assumption that guides me is that exceptional storylines of extraction exert epistemic violence in that they dismiss, marginalise, and silence its associated colonial and environmental harm past and present. Complicity in such violence allows them to uphold control over the land and resources of the North (Quijano 2000). In what follows, I identify tendencies of extractive exceptionalism in the data as two-fold. First, exceptional resource richness provides Sweden with an opportunity to contribute to the international market and care for the Swedish people. Second, Swedish mining is imagined to be exceptionally non-intrusive, and environmentally and socially sustainable.

The story of Swedish mining as told by pro-mining voices is one of progress, growth, and opportunity. The mining industry has a commonsensical status as an engine for growth, employment, regional and national development, as a source of prosperity, social and environmental benefit. Former Prime Minister Stefan Löfven (resigned in November 2021) frequently reiterated the need for mining not only for employment and economic growth, but also for advancing Sweden's global environmental leadership (NSD, September 24, 2021). It is commonly stressed that Swedish iron ore is an asset also to the international community where Sweden contributes with about 90 per cent of Europe's total iron ore extraction, invoking its prominence beyond its borders. In regard to critical raw materials, Sweden's role within the EU has increased even further with the recent EU push for self-sufficiency. Now, as is stated by the Swedish Association of Mines, Minerals and Metal Producers, "the spotlight is on Sweden, which is already today Europe's most mining nation (SveMin, September 4, 2020)." Similarly, for Löfven:

> Since we are gifted with these assets, we should of course make use of them to the greatest extent possible.
>
> (own translation, Löfven 2019)

Swedish resource abundance is an asset, a gift that Sweden logically ought to embrace. Once extracted, "Sweden's iron ore deposits will do the world good" (Mining for Generations, accessed March 25, 2021). This is communicated on the website of the Swedish government and business initiative *Mining for generations—Mining by Sweden*, a communication platform aimed to boost the image of Swedish mining and invite cross-country collaborations. The initiative presents Sweden as a country that welcomes investment and stresses the richness of "our" bedrocks and resource deposits, "just waiting for future exploration and development"—indeed, a "hard rock paradise" for investors (Mining for Generations, accessed March 25, 2021). Mining is narrated as an undertaking rooted in national tradition and custom, with the knowledge accumulated throughout the country's long mining history making it an attractive destination for investment. History is also invoked in the telling of expansion to the North. The discoveries in the North are on the website presented as having incited hope in the form of a "new future," a way out of poverty and famine during economic hardship.

> At a time when 25% of Swedes were immigrating to America in an attempt to avoid starvation and poverty, the need for this new future was apparent.
> (Mining for Generations, accessed March 25, 2021)

What was imagined as the "land of the future" allowed a poverty-ridden Sweden to emerge as a modern welfare state (Sörlin 1988). Such visions of the North as an unseized frontier of opportunity, empty and void of bodily obstacles, grants a path out of despair yet neglects the force with which advancements were historically instigated. The expansion through Norrland and the ease with which property over "our" discovered deposits was and continues to be claimed tends to render invisible the Sami population and perspectives. It displays an example of the forgetfulness with which Swedish extractive expansion is cemented in the collective remembering, presented purely as an opportunity, a lifeline for the Swedish people. The notion of emptiness emulates the wider story told about the Arctic as vacant space (Bergman Rosamond and Rosamond 2015; Lindberg 2019), as unexplored, unexploited and thus full of potential, profitable and transformable, a no man's land waiting to be discovered and put to proper use.

British Beowulf Mining also leans on the colonial/extractive episteme when legitimising land investments in Sweden, imagining Kallak as empty and void of obstacles, as *terra nullius* in Sweden's midst. Its presentation material illustrates that "Kallak is a quality magnetite iron ore deposit" and thereby a "real asset." The company presents a timeline for the "Story of Kallak," which starts with its "discovery" by SGU in 1947 and continues with the announcement of Kallak being the "largest unexploited iron ore deposit in Sweden" in 1970 (Beowulf Mining 2017). Anything prior to that discovery, before its assertion of value, is irrelevant and thus excluded. With similar tendencies of the state imaginary of the North, associating Kallak's materialisation with the discovery of its commercial value—or what Arturo Escobar (1999) refers to as a "capitalist-nature" ontology—exhibits connotations of settler mentality, an unawareness and indifference to what has been prior or that which exists alongside ore deposits. In this way, Kallak's timeline silences relationalities to land that are unrelated to mining, making the story of Kallak purely one of resources, of profit, iron ore, and extraction.

Mining is discursively associated with employment opportunities, an increased standard of living, and a revival of otherwise digressing municipal districts. With this, expansion through the North is understood as an opportunity to develop Sweden's sparsely populated regions. According to Beowulf Mining, Kallak is "a real opportunity to transform Jokkmokk." Kallak has, if managed appropriately, grand "potential" to create direct and indirect employment, generate tax revenues, stimulate the economy, and make Jokkmokk an attractive site for development. Then chairman of Jokkmokk's municipal council Robert Bernhardsson fears that "local growth and investment climate is moving towards a new ice age. If this continues, Jokkmokk will freeze to death" (Beowulf Mining 2017). When asked about protests against the initiation of the mine, he also stresses that, "I think it is actually terrible that people stand in the way of hundreds of jobs in a sparsely populated municipality like Jokkmokk"[7] (Sveriges Radio, August 8, 2020). We encounter here a story of a currently underperforming region that is walking down a frosty path of destructive demographics, out-migration, and loss of workforce. Yet the story also opens a door to a hopeful future, in which Jokkmokk can be brought onto the right path by opening its doors to mining. The storyline of exceptional deposits displays clear binaries—between opportunities seized and lost, between transformation and a lingering in a destructive status quo, between hope and despair, and indeed between life and death. These binaries propose a rational alignment with the former rather than the latter, conditioning hopeful notions of development, employment, and social welfare with the presence of mining, its absence with dismay. Not aligning is narrated as "terrible," as irrational, incompatible with the project of modernity (Quijano 2000), thereby rationalising and rendering logical a commodification of the land of the North.

In addition to this first avenue of exceptionalism in which the North is constructed as an abundant resource pool of opportunity, Sápmi is also habitually constructed as an area in which multiple diverse interests can coexist harmoniously. The Swedish mineral strategy envisages the expansion of the Swedish mining industry in harmony with other activities, including reindeer herding, tourism, and environmental initiatives (Government Offices of Sweden 2013, 23). Given its assumed emptiness, it is commonly argued that there is room for mining in Sweden's second biggest municipal district without posing a risk to local culture and heritage (see also Ojala and Nordin 2019, 114). Growth and prosperity are envisioned in the synergy of these diverse activities. Sweden has the resource capacity, technical expertise, legislation, and respect for environmental, social, and cultural sustainability that makes it a sound choice to situate the extraction of necessary resources in its midst. The alternative to the expansion of mining in Sweden is presented as one in which "child labour, polluted rivers or exploited workers" (Baylan 2020) may await, making the Swedish expansion of the mining sector indeed the responsible choice. Swedish mining is also imagined non-intrusive in that it is discursively disassociated from environmental degradation and purely presented as a source of development and progress, stressing the necessity of extraction for the modern way of life (Government Offices of Sweden 2013).

Figure 4.1 Den Svenska Gruvan 2021, www.densvenskagruvan.se (own website stillframe).[8]

Illustrated above, the joint marketing campaign of Swedish mining companies, *The Swedish Mine* (Den Svenska Gruvan), highlights the everyday benefits associated with metals extraction. It presents mining as a prerequisite for making "our modern and sustainable lives possible" (Figure 4.1).

The notion of non-intrusion and harmonious coexistence allows for carefree, comfortable dwelling in the forgetful reproduction of the colonial/extractive episteme. We now turn to the disruption of such forgetfulness through an analysis of instances of *decolonial recollection*.

Decolonial recollection: challenging the virtue of extraction

As we have seen, pro-mining actors reproduce the colonial/extractive episteme when emphasising Swedish mining exceptionalism. Conversely, anti-mining voices highlight silences in such exceptionalism. I propose the concept of *decolonial recollection* to provide a reading of discourses that disrupt the linearity of assumed extractive exceptionalism as they point to its forgetful tendencies.

Narratives of the exceptionalism of Swedish resource deposits display propertarian language that allows for an unproblematic expansion through the North, in which "we" must make use of "our" resources that "we are gifted with" (Löfven 2019). As we shall see, decolonial recollections construct Swedish expansion instead as theft of land, of colonisation, forced eviction, non-recognition, violent intrusion, racial discrimination, and neglect. Instead of narrating Norrland as an unseized opportunity void of obstacles for Swedish discoveries, the violent nature of such expansion is recollected in attempts to challenge the naturalised idea of Sweden's rightful claim to the North. For instance, musician and activist Sofia Jannok rejects the Swedish claim to ownership over Sápmi in her song *This is my land*, in which she wanders with a Sami flag gazing over the land that she sees as hers/theirs with the camera recording through the flag, colouring Sápmi in the colours of its peoples (Jannok 2016). Instead of the pro-mining emphasis on coexistence and harmony between diverse activities, anti-mining disruptions stress the need to prioritise between mutually exclusive land use. The story of mining as cohabiting Sápmi side-by-side and with respect for Sami tradition and culture is rejected as unrealistic. Instead, when it comes to mining "the question is, at the expense of what? At the expense of whom?" (own translation, Märak 2015) The mine is here no longer narrated as a life-giving entity rich of opportunity, but instead, as a symbol of demise. Reindeer husbandry, ecotourism, and other traditional and local industries are here presented as the backbone of Jokkmokk's economy and stability, challenging the idea of a dying Jokkmokk that must be rescued by extraction (Märak 2015). Instead of municipal survival being dependent on mining, diversified livelihood opportunities are understood as threatened.

While municipal death is rejected, notions of death and a fight for survival are also reoccurring narrative tools in decolonial recollection. The rebel group *tjáhppis rájddo* initiated by the Sami activist sisters Mimi and Maxida Märak marched during the traditional Jokkmokk winter market as a critique against a selective interest in Sami peoples as a lucrative source of revenue for the Swedish tourism industry (Märak 2015). Participants marched while dressed in black waste bags, faces painted white and reindeer corpses pulled after them in sleds. The march connotes a dying people, the waste bags insinuate a gaze from the outside onto a people that are disregarded, disposable, yet—despite the threat to their existence—are determined to walk on in the tracks of those that marched before them (Figure 4.2).

Mimi Märak addresses mining intrusion also in her performance of a poetry-slam piece in Gállok in 2013. She tells a story of a world scenic and pristine, of cherished land and white snow left dead and tainted after its encounter with extractive machines. Märak associates the misappropriation of land and flora with the suffocation of the Sami peoples. She presents attempts to divorce the Sami from their land as a threat to cultural survival, though efforts to displace them from their lands would be fruitless as their

Figure 4.2 Preparations for Tjáhppis Rájddo manifestation during the 2014 Jokkmokk winter market.

(Photo: Tor Lennart Tuorda)

ties go too deep to be uprooted, inferring a union, a merger with nature. She is one with the land, her legs the roots that anchor her, the suffocation of the land resulting in the suffocation of her peoples.[9] In this way, the unexploited land becomes here a symbol for life rather than death. The analogy of roots also a reference to the long tradition of Sami presence, challenges the idea that the Sami as traditionally nomadic can readily be relocated for the benefit of mining.

As we have seen above, the emptiness and invisibility that is imposed through the imaginary of extractive exceptionalism reinforces Swedish colonial innocence and legitimises further expansion. Decolonial interceptions of acclaimed emptiness are numerous, and Beowulf Mining chairman Sinclair-Poulton's "What local people?" has further inspired discursive avenues of decolonial recollection in the case of Gállok. The response "us local people," "we are the local people," attempts to "bust the myth of non-existent locals, and to showcase the growing opposition to the mining industry's environmental and cultural destruction" (Figure 4.3).[10]

The photo below is the same that Sinclair-Poulton pointed to when suggesting emptiness but is here reappropriated as a backdrop by the "What local people" Facebook community, thereby reclaiming and occupying that which has been rendered vacant in stories of exceptionalism. The Facebook page also presents a collection of close-up head shots of locals assembled to function as the group's profile picture, suggesting a collected, unified identity, looking directly into the camera lens, insisting on being seen and forging recognition where there was none granted. In this decolonial recollection, the image of the seemingly vacant landscape aids in telling a different story, not one of emptiness and absence but indeed one of unyielding presence. Similarly, Mimi Märak enacts Gállok as filled with and occupied by locals ready and waiting to burst to the forefront to protect the land from any attempts to alter and destroy it.[11]

Figure 4.3 Profile picture and community background image of the *What local people* Facebook community page (stillframe accessed March 3, 2022).

Figure 4.4 Anders Sunna's art installation, used as the focal point in Sofia Jannok's music video *We are still here* in 2016 in Gállok (anderssunna.com).

Resistance artists Anders Sunna and Sofia Jannok also invoke presence in their joint art installation, *We are still here*. It displays the iconic Sami activist Elsa Laula Renberg alongside new generations in a reassurance of continued resistance. This disruption of acclaimed emptiness, vastness, and harmonious non-intrusion in Sápmi seeks recognition of continued connectivity to the land while placing the fight for Gállok into a historic context of enduring rights violations. In so doing, it reminds the Swedish extractive exceptionalism of the violence that is exerted in the reproduction of the colonial/extractive episteme (Figure 4.4).

Conclusion

In this chapter, I have proposed the concept of *decolonial recollection* to help grasp the subversive power that lies in disruptions of the forgetful imaginaries of exceptionalism. I have argued that the narratives of extraction and the Swedish virtue of mining that I have illustrated through the case of Gállok/Kallak allow for comfortable dwelling in Swedish colonial forgetfulness (Maldonado-Torres 2004), and thereby for Sweden's mining identity to coexist alongside its good state (Lawler 2013) persona, while dodging extensive need for self-scrutiny. The narratives of the empty land of the North, the historic importance of its bedrocks, and Sweden's non-intrusive extraction practices are deeply anchored in and reinforce a sense of self as virtuous in various ways, not least as pertaining to colonial innocence. Such narratives effectively legitimise further extractive capitalist investments in land in Sápmi. Meanwhile, colonial ties, land enclosures, local interests, and environmental harms are sidelined, dismissed, irrationalised in the quest for that which is understood as progress. This form of epistemic violence (Grosfoguel 2013; Spivak 2010) protects the stability and coherence of the idea of mining as virtuous, and the dominant episteme upon which it rests against disruptions. Still, and this lies at the heart of this chapter, Sami activists have forcefully tapped on the shoulder of Swedish extractive exceptionalism by re-collecting indigenous perspectives that have been swept under its protective rug. Decolonial recollections function to disrupt the strategically negligent colonial/extractive episteme and provoke self-scrutiny at least to the extent that requires a response that reinstates the stability of acclaimed exceptionalism. In the time of writing, the government decided to award Beowulf Mining an exploitation permit; however, resistance is not admitting defeat. It is due to the unresolved status of this project beyond the scope of this chapter to assess the potential of decolonial recollection to trigger sustained change. Still, I interpret the prolonged permit process in an otherwise endorsed industry, the heightened media visibility, government attempts to address critiques (not least by introducing unprecedented conditions to the granted permit), and increased engagement of the general public, as manifesting provoked self-scrutiny of the virtuous mining nation.

In this way, the chapter also contributes to the debate on exceptionalism with an analysis of the imaginary of Swedish mining as virtuous, responsible, innocent. While the Swedish sense of self has been dissected in regards to its supposed colonial neutrality (Fur 2013; Keskinen et al. 2009; Ojala and Nordin 2019) and colour-blindness (Habel 2012; Hübinette 2013), I have pointed to Swedish extractive activities as an additional avenue of exceptionalism. I have argued that the story of extractive exceptionalism—a nation gifted with abundant resource deposits that extracts without intruding—is a strategically incomplete imaginary of past and present activities in the North. Swedish mining is constructed as equivalent with growth, opportunity, a source of competitiveness, neighbourliness, harmony, and coexistence. It is a linear storyline, one beyond rational questioning, taken for granted in light of the good that Swedish mining can do for its people. It would be illogical not to make proper use of "our" resources that "we" are "gifted with," through which "we" will blossom, and which allow "us" to live a carefree, modern life. With propertarian language, resources of the vast and empty North are subsumed under the realm of the Crown, continuously envisioned as a sphere of opportunity, a "paradise" to be seized. While discursively disassociating mining from any form of intrusion—may it be environmental, cultural, or colonial—it is presented as a prerequisite for modernity. Modernity and green growth are constructed as conditioned by the presence of mining, failing to embrace the mine means standing in the way of progress. As such, not aligning with "paradise" is rendered irrational, "terrible." The story of extractive exceptionalism is controlled through its many silences. It is forgetful (Maldonado-Torres 2004) of the Sami people and the violence through which expansion was made possible, and also of the environmental harm that extraction can trigger. The idea of harmonious coexistence disregards the exclusion that mining requires. The misrecognition of Sami presence reproduces a logic of conditionality of legitimate land use tied to progress and efficiency, making visible only that which aligns with the utilitarian "capitalist-nature" regime (Escobar 1999).

Normalising such sentiments of virtuous extraction allows for mining expansion to be placed out of scrutiny's reach. However, as illustrated through the case of Gállok, the consistency of extractive exceptionalism is forcefully disrupted in instances of decolonial recollection. These are retaliations (Galván-Álvarez 2010) against the violence exerted through the reproduction of the dominant colonial/extractive episteme, and against the "coloniality of power's" stubborn grip (Quijano 2000). That which is kept in the dark to allow for a linear stride of envisioned exceptionalism, is here recollected to reinscribe an association of mining with destruction, colonial violence, demise, exclusion. Out of the shadows of exceptionalism's neglect step bodies and experiences that demand to be seen, that remind the "heads and hearts of people" (Wekker 2016, 19) in which the undoubtedly good and virtuous nature of mining is enshrined, that "We are still here."

Notes

1 Original Swedish: "Vår gruvnäring och vårt järnmalm är ju för oss vad oljan är för norrmännen. En fantastisk rikedom en möjlighet att bygga framtidsinvesteringar, framtidsutveckling [...]".
2 Original Swedish: "Stål har byggt Sverige, och stål har byggt vår välfärd."
3 The Sami and Swedish name, respectively.
4 International Labour Organization's convention No. 169 (or the Indigenous and Tribal Peoples Convention) is a binding convention that among other things recognises indigenous peoples' rights to ownership of traditionally occupied land and the right to be consulted for extraction objectives of sub-surface resources (Article 14-15, ILO 1989).
5 Geological Survey of Sweden (Swedish: Sveriges Geologiska Undersökning).
6 Konstitutionsutskottet (KU), oversees and ensures that the government respects and follows existing regulations and rules. As a tool of parliamentary control, members of the Riksdag can report government processes and ministers for scrutiny by the Committee (Government Offices of Sweden, accessed February 2022).
7 Original Swedish: "[...] jag tycker att det är faktiskt fruktansvärt att man på det här sättet motverkar hundratals arbetstillfällen i en glesbygdskommun som Jokkmokk." https://sverigesradio.se/artikel/7529296.
8 Own translation of the text in the image: "Metals and minerals make our modern and sustainable lives possible. The commute to work, the zipper in the jacket, the bridge over the road, your smart phone, the battery in the car, the wind power park, the button in trousers and the solar cell on the roof—everything starts under the ground. In Sweden, we are building the world's most climate-smart mining industry. Welcome to the Swedish Mine."
9 [...] Vi som älskar och önskar att våran grönska ska få vara kvar där den är utan att ni skär den tär den, för faktum är, att ni stryper en minoritet. Vet att vi står med ena foten i graven, men roten går för djupt för att slita loss. (excerpt from Märak 2013).

Own translation: Us who love and want to see our greenery remain where it is without you cutting it, exhausting it, because the truth is, that you are strangling a minority. Know that we are standing with one foot in the grave, but the root goes too deep to be torn out.

10 *Own translation*, original Swedish: "För att slå hål på myten om den obefintliga lokalbefolkningen och visa att det finns ett utbrett motstånd mot den miljö- och kulturförstörelse som gruvexploateringarna innebär finns denna sidan." ('What local people' Facebook community page, accessed March 3, 2022).
11 Original Swedish: [...] De ska skämmas, stämmas för varje damm som skulle dämmas varje träd du vill fälla, från varje plats du vill spränga, kommer det tränga fram tusen röster från oss som är What local people (excerpt from Märak 2013).

Own translation: They shall feel shame, be prosecuted for every dam that was going to be built, every tree you want to cut down, from every place you want to blow up, there will emerge a thousand voices from us who are What local people.

References

Baylan, Ibrahim. 2020. *Interpellation 2020/21:216*. Video recording. https://www.riksdagen.se/sv/webb-tv/video/interpellationsdebatt/tillstandsprovningen-gallande-kallak-och-laver_H810216

Beowulf Mining. 2017. *KALLAK: A Real Asset, and a Real Opportunity to Transform Jokkmokk*. Copenhagen Economics.

Bergman Rosamond, Annika. 2020. "Music, Mining and Colonisation: Sámi Contestations of Sweden's Self-narrative." *Politik* 1 (23): 70–86.

Bergman Rosamond, Annika, and Ben Rosamond. 2015. "New Political Community and Governance at the Top of the World: Spatiality, Affinity and Security in the Arctic." In *Governing Borders and Security: The Politics of Connectivity and Dispersal*, edited by Ted Svensson and Catarina Kinnvall, 135–52. Abingdon: Routledge.

Brunner, Claudia. 2020. *Epistemische Gewalt: Wissen und Herrschaft in der kolonialen Moderne*. Edition Politik. Bielefeld: Transcript Verlag.

Brysk, Alison. 2009. *Global Good Samaritans: Human Rights as Foreign Policy*. Oxford: Oxford University Press.

Butler, Judith. 2010. *Frames of War: When is Life Grievable?* London: Verso.

De la Cadena, Marisol, and Mario Blaser. 2018. *A World of Many Worlds*. Durham: Duke University Press.

de Sousa Santos, Boaventura. 2014. *Epistemologies of the South: Justice Against Epistemicide*. Boulder: Paradigm Publishers.

Donadey, Anne. 1999. "Between Amnesia and Anamnesis: Re-membering the Fractures of Colonial History." *Studies in 20th & 21st Century Literature* 23 (1): 111–116.

Drugge, Anna-Lill, ed. 2016. *Ethics in Indigenous Research: Past Experiences: Future Challenges*. Umeå: Vaartoe/Centre for Sami Research (CeSam).

Escobar, Arturo. 1999. "After Nature: Steps to an Antiessentialist Political Ecology." *Current Anthropology* 40 (1): 1–30.

Escobar, Arturo. 2018. *Designs for the Pluriverse: Radical Interdependence, Autonomy, and the Making of Worlds. New Ecologies for the Twenty-First Century*. Durham: Duke University Press.

Fanon, Frantz. 2001. *The Wretched of the Earth*. London: Penguin Books.

Fraser Institute. 2020. "Annual Survey of Mining Companies, 2019." February 25, 2020. https://www.fraserinstitute.org/studies/annual-survey-of-mining-companies-2019.

Fur, Gunlög. 2013. "Colonialism and Swedish history: Unthinkable connections?" In *Scandinavian Colonialism and the Rise of Modernity*, edited by Magdalena Naum and Jonas Nordin, 17–36. New York: Springer.

Galván-Álvarez, Enrique. 2010. "Epistemic Violence and Retaliation: The Issue of Knowledges in Mother India." *Atlantis. Journal of the Spanish Association of Anglo-American Studies* 32 (2): 11–26.

Gómez-Barris, Macarena. 2017. *The Extractive Zone: Social Ecologies and Decolonial Perspectives*. Durham: Duke University Press.

Government Offices of Sweden. 2013. *"Sveriges Mineralstrategi: För ett hållbart nyttjande av Sveriges mineraltillgångar som skapar tillväxt i hela landet." N2013.02.*

Grosfoguel, Ramón. 2013. "The Structure of Knowledge in Westernized Universities." *Human Architecture: Journal of the Sociology of Self-Knowledge* 11 (1): 73–90.
Habel, Ylva. 2012. "Challenging Swedish Exceptionalism? Teaching While Black." In *Education in the Black Diaspora - Perspectives, Challenges, and Prospects*, edited by Kassie Freeman and Ethan Johnson, 99–122. London: Routledge.
Harvey, David. 1996. *Justice, Nature and the Geography of Difference*. Oxford: Blackwell.
Hernández Castillo, Aída. 2020. "'Putting Heart' into History and Memory: Dialogues with Maya-Tseltal Philosopher, Xuno López Intzin." *Memory Studies* 13 (5): 805–819.
Hjälmered, Lars. 2020. *Interpellationsdebatt - Interpellation 2020/21:216*. Video Recording. https://www.riksdagen.se/sv/webb-tv/video/interpellationsdebatt/tillstandsprovningen-gallande-kallak-och-laver_H810216
Hübinette, Tobias. 2013. "Swedish Antiracism and White Melancholia: Racial Words in a Post-Racial Society." *Ethnicity and Race in a Changing World* 4 (1): 24–33.
Jannok, Sofia. 2016. *This is my Land*. Music Video. https://youtu.be/riXVuhlMNQA
Keskinen, Suvi, Salla Tuori, Sari Irni, and Diana Mulinari, eds. 2009. *Complying with Colonialism: Gender, Race and Ethnicity in the Nordic Region*. Aldershot: Ashgate.
Körber. 2019. "Sweden and St. Barthélemy: Exceptionalisms, Whiteness, and the Disappearance of Slavery from Colonial History." *Scandinavian Studies* 91 (1–2): 74–97.
KU, Committee on the Constitution. 2020. *Konstitutionsutskottets Betänkande: Granskningsbetänkande Hösten 2020. 2020/21:KU10*.
Lander, Edgardo. 2002. "Eurocentrism, Modern Knowledges, and the 'Natural' Order of Global Capital." *Nepantla: Views from South* 3 (2): 39–64.
Lawler, Peter. 2013. "The 'Good State' Debate in International Relations." *International Politics* 50 (1): 18–37.
Lawrence, Rebecca. 2014. "Internal Colonisation and Indigenous Resource Sovereignty: Wind Power Developments on Traditional Saami Lands." *Environment and Planning D: Society and Space* 32 (6): 1036–53.
Lawrence, Rebecca, and Mattias Åhrén. 2017. "Mining as Colonisation: The Need for Restorative Justice and Restitution of Traditional Sami Lands." In *Nature, Temporality and Environmental Management: Scandinavian and Australian Perspectives on Peoples and Landscapes*, edited by Lesley Head, Katarina Saltzmann, Gunhild Setten, and Marie Stenseke, 149–66. Abingdon, Oxon: Routledge.
Lindberg, Helena Gonzales. 2019. *The Constitutive Power of Maps*. Doctoral Dissertation, Lund: Lund University.
Lindmark, Daniel. 2013. "Colonial Encounter in Early Modern Sápmi." In *Scandinavian Colonialism and the Rise of Modernity*, edited by Magdalena Naum and Jonas Nordin, 131–146. New York: Springer.
Löfven, Stefan. 2019. "Fortsatt Beslutsvånda Om Laver Och Kallak." *Sveriges Radio P4 Norrbotten*. https://sverigesradio.se/artikel/7164962
Löfven, Stefan. 2020. "Tal Av Statsminister Stefan Löfven På Invigningen Av HYBRIT:S Nya Pilotanläggning." Regeringskansliet. August 31, 2020. https://www.regeringen.se/tal/2020/08/tal-av-statsminister-stefan-lofven-pa-invigningen-av-hybrits-nya-pilotanlaggning/
Maldonado-Torres, Nelson. 2004. "The Topology of Being and the Geopolitics of Knowledge: Modernity, Empire, Coloniality." *City* 8 (1): 29–56.
Märak, Mimi. 2013. *What Local People? Poesi För Ett Gruvfritt Kallak*. Video recording. https://youtu.be/JiFcEvjIG8w

Märak, Maxida. 2015. *Maxida Märak*. SverigesRadio: Sommar & Vinter i P1. Radio recording. https://sverigesradio.se/sida/avsnitt/583046?programid=2071

Merchant, Carolyn. 1980. *The Death of Nature: Women, Ecology, and the Scientific Revolution*. 1st ed. San Francisco: Harper & Row.

Mignolo, Walter D. 2007. "Delinking: The Rhetoric of Modernity, the Logic of Coloniality and the Grammar of De-coloniality." *Cultural Studies* 21 (2–3): 449–514.

Mignolo, Walter. 2011. *The Darker Side of Western Modernity: Global Futures, Decolonial Options*. Durham: Duke University Press.

Mörkenstam, Ulf. 2019. "Organised Hypocrisy? The Implementation of the International Indigenous Rights Regime in Sweden." *The International Journal of Human Rights* 23 (10): 1718–1741.

Mulinari, Diana, Suvi Keskinen, Sari Irni, and Salla Tuori. 2009. "Introduction: Postcolonialism and the Nordic Models of Welfare and Gender." In *Complying with Colonialism: Gender, Race and Ethnicity in the Nordic Region*, edited by Suvi Keskinen, Salla Tuori, Sari Irni, and Diana Mulinari, 1–16. Aldershot: Ashgate.

NSD. 2021. "Löfven Klargör S-Linjen: Fler Gruvor Ska Öppnas." Norrländska Socialdemokraten, NSD. September 24, 2021. https://nsd.se/artikel/jv9v4d7l.

Ojala, Carl-Gösta, and Jonas Nordin. 2019. "Mapping Land and People in the North: Early Modern Colonial Expansion, Exploitation, and Knowledge." *Scandinavian Studies* 91 (1–2): 98.

Össbo, Åsa, and Patrik Lantto. 2011. "Colonial Tutelage and Industrial Colonialism: Reindeer Husbandry and Early 20th-Century Hydroelectric Development in Sweden." *Scandinavian Journal of History* 36 (3): 324–48.

Patterson, Molly, and Kristen Renwick Monroe. 1998. "Narrative in Political Science." *Annual Review of Political Science* 1 (1): 315–31.

Quijano, Aníbal. 2000. "Coloniality of Power, Eurocentrism, and Latin America." *Nepantla: Views from South* 1 (3): 533–80.

Rose, Gillian. 2001. *Visual Methodologies : An Introduction to the Interpretation of Visual Materials*. London: SAGE.

Sandström, Moa. 2020. *Dekoloniseringskonst: Artivism i 2010-talets Sápmi*. Doctoral Dissertation, Umeå: Umeå University.

Sehlin, MacNeil Kristina. 2017. *Extractive Violence on Indigenous Country: Sami and Aboriginal Views on Conflicts and Power Relations with Extractive Industries*. Doctoral Dissertation, Umeå: Umeå university.

Sinclair-Poulton, Clive. 2012. Beowulf Chairman Talks About "superb" Drilling Results. Videorecording. http://ravarumarknaden.se/intressant-framtid-for-beowulf-mining/

Sörlin, Sverker. 1988. *Framtidslandet: Debatten Om Norrland Och Naturresurserna under Det Industriella Genombrottet*. Doctoral Dissertation, Umeå: Umeå university.

Spivak, Gayatri Chakravorty. 2010. "Can the Subaltern Speak?" In *Can the Subaltern Speak? Reflections on the History of an Idea*, edited by Rosalind C. Morris, 24–28. New York: Colombia University Press.

Svemin, the Swedish Association of Mines, Mineral and Metal Producers. 2020. "New EU Initiative Increases Demand for Swedish Mines." September 4, 2020. https://www.svemin.se/en/news/news/new-eu-initiative-increases-demand-for-swedish-mines/.

Sveriges Radio. 2012. "Miljarderna Försvinner - Skiten Blir Kvar." October 26, 2012. https://sverigesradio.se/artikel/5325964.

SVT. 2022. "Näringsministern Om Gruvtillstånd: "Skaver Ofta Mellan Företag Och Länsstyrelsen"." SVT News. January 23, 2022. https://www.svt.se/nyheter/lokalt/

norrbotten/naringsministern-det-skaver-ganska-ofta-mellan-foretag-och-lansstyrelsen.

Swedish Ministry of Justice. 1986. *SOU 19886:36 Samernas folkrättsliga ställning*. Stockholm: Liber Tryck AB.

The Decolonial Atlas. n.d. "Sápmi: The Sámi Homelands." The Decolonial Atlas. Accessed December 8, 2022. https://decolonialatlas.wordpress.com/2021/02/08/sapmi-the-sami-homelands/.

Thiong'o, wa Ngũgĩ. 1987. *Decolonizing the Mind: The Politics of Language in African Literature*. Harare: Zimbabwe Publishing House.

Tlostanova, Madina, and Walter Mignolo. 2012. *Learning to Unlearn: Decolonial Reflections from Eurasia and the Americas*. Columbus: Ohio State University Press.

Tsing, Anna Lowenhaupt. 2015. *The Mushroom at the End of the World: On the Possibility of Life in Capitalist Ruins*. Princeton: Princeton University Press.

Tuorda, Tor Lennart. 2014. *What Local People?* Video recording Video recording. https://youtu.be/XIvmbifyQ80

Vuorela, Ulla. 2009. "Colonial Complicity: The 'Postcolonial' in a Nordic Context." In *Complying with Colonialism: Gender, Race and Ethnicity in the Nordic Region*, edited by Suvi Keskinen, Salla Tuori, Sari Irni, and Diana Mulinari, 19–33. Aldershot: Ashgate.

Wekker, Gloria. 2016. *White Innocence: Paradoxes of Colonialism and Race*. Durham: Duke University Press.

5 Coloniality of knowledge and the responsibility to teach

Nordic educational interventions in the "South"

Jelena Vićentić

Introduction

This chapter addresses the project of global education as it is configured in multilateral development institutions, paying particular attention to the role of the Nordic actors in this regard. While framed as "global," such educational initiatives are, however, in practice directed towards the "South," and appear as projects that present peoples, communities, and individuals of the South with the continuous demand for "improvement," adjustment, and realignment to modernity/coloniality[1] by demanding assimilation to worldviews and value systems that perpetuate capitalism, patriarchy, Eurocentrism, and racism. This dynamic occurs between the reproduction of narratives, methodologies, and hierarchic categories conducive to the preservation of coloniality, "the most general form of domination in the world today, once colonialism as an explicit political order was destroyed" (Quijano 2007, 170). It occurs through erasures of the epistemologies that challenge the *status quo*.

Regarding the "confluence" of the European colonial domination and the formation of the European modernity/rationality, Aníbal Quijano points to the concurrent processes as "producing a universal paradigm of knowledge and of the relation between humanity and the rest of the world" (2007, 172). Walter Mignolo subsequently identifies salvation, progress, development, modernisation, and democracy as expressions of the rhetoric of modernity reproducing and perpetuating the logic of coloniality (2011, 13). As successive and cumulative stages, these concepts are historically linked to the promise of salvation through Christianity, continuing with the civilising mission and the bio-politics of the state control, and finally, at the current stage, they enhance the logic of development and the promise of the market (Mignolo 2011, 13).

The close connection between "official" knowledge and colonial power is reflected in the predominant imaginary, which assumes "scientific knowledge to be the knowledge produced by white, male, property-owning, (culturally) Christian elites of a few countries in the world, whose 'science' in turn rest upon the extraction, appropriation, disavowal, or invisibilisation of all other existing knowledges" (Finck-Carrales & Suárez-Krabbe 2022). According to Walter Mignolo and Catherine Walsh, knowledge occurs as the paramount

DOI: 10.4324/9781003293323-6

domain of the colonial matrix of power, while economy is "knowledge organizing and legitimizing praxis" (2018, 177). Education thus appears as an intermediary step between the two, enabling the perpetuation of the coloniality of knowledge as its tool of dissemination:

> Modern (European) civilization understands itself as the most developed, the superior, civilization; This sense of superiority obliges it, in the form of a categorical imperative, as it were, to 'develop' (civilize, uplift, educate) the more primitive, barbarous, underdeveloped civilizations.
> (Dussel 1993, 75)

Due to the complex and historically burdensome legacy of the aid to education in the Global South, this chapter focuses on initiatives within this field offered by Nordic actors.[2] In this regard, I elaborate on two readings of global education which warrant further study. Firstly, that the project of "assisted" global education is a foundational element of the development doctrine and of constitutive importance for the preservation of coloniality. Assistance to education is also a part of the neoliberal development interventions conducted by a range of actors, including international bodies, corporate entities, governments, and non-governmental organisations. Education is presented as the key element in increasing economic competitiveness and portrayed as melding individual and collective economic benefits. Simultaneously, it is supposed to be assisted through the establishment of common standards, while being exploited as an economic opportunity. The requirement of curricular homogeneity expands the market for "universal" school materials. Additionally, "not-in-school" children present an untapped resource and an opportunity for private education providers.[3] These practices are consistent with the critical view of the neoliberal globalising interventions as (re)colonisation, "embedding and re-embedding of neo-liberalism utilizing multiple avenues including institutional, state, corporate and intellectual pressure" (Bargh cited in Choudry 2009, 96).

Secondly, a decolonial reading of programmatic documentation of global education initiatives illuminates its many incongruities. Here the educational intentions presented as benevolent collide with colonial continuities. The contradiction makes erasures and assigned meanings become visible through the decolonial lens. The modern global order rests on exclusions from the province of knowledge, while "undermining the other's capacity as a knower in principle also fundamentally undermines such person as a human being" (Mungwini 2017, cited in Manthalu and Waghid 2019, 28).

The following section introduces pertinent aspects of the many overlapping initiatives within the global education project, including the *Education for All* movement, the *Global Partnership for Education*, the *Education Commission*, *Education Outcomes Fund, Education Cannot Wait*, as well as a recent response to the COVID-19 pandemic the *Save Our Future* campaign (Benavot et al. 2015; ECW 2018; Save Our Future 2020). It also outlines the role of Nordic states as initiators, promoters, and donors in this thematic field (GPE 2020, 122; Hermann 2018; NORAD 2017c; UNESCO 2019). Subsequently, these initiatives are explored within meta-narratives of development, including the notion of development as a gift and a projection of ideal or idealised institutions, with "universal" values set as criteria for successful transfer and result evaluation. Some of the key elements of the global education narrative are examined, including the alarmist notion of the "learning crisis" and the historical and geographical (un)rootedness of the project. This is a space of (dis)connection between the donor or the education aid platform on one side, and the recipient—states, communities, and "beneficiaries"—on the other. A decolonial reading of the constructed spaces of intervention opens the possibility of understanding colonial continuities and workings of the education project within the colonial matrix of power. In particular, the absence of the agency of the recipient state and discursive naturalisation of deprivations is implicit in those spaces: hegemonic neoliberal forms of production of globalisation are naturalised and counterhegemonic forms of agency made invisible (Suárez-Krabbe 2012, 33). Empowerment is projected as being externally induced, especially for the essentialised "girl child" of the South (Dogra 2012).

Global education initiatives

The rise of the global education project can be traced to the World Declaration on Education for All (EFA) adopted in 1990 in Jomtien, Thailand, which reaffirmed education as a basic human right, as proclaimed in the Universal Declaration of Human Rights. The event presents the point of departure for international measures targeting global education. Interventionist from its inception, the EFA rests upon the conviction that "public policy can radically transform education systems and their relationship to society within a few years, given adequate political will and resources" (UNESCO 2008, cited in Benavot et al. 2016, 241). Notably, the goals of the EFA include improving comprehensive early childhood care and education; access to free and compulsory primary education of good quality for all children, particularly girls, children in difficult circumstances, and those belonging to ethnic minorities (by 2015); access to appropriate learning and life-skills programmes; 50 per cent improvement in adult literacy and access to basic and continuing education (also by 2015); eliminating gender disparities in primary and secondary education by 2005 and achieving gender equality in education (also by 2015); and improving the quality of education.

EFA was adopted as a programme under the UNESCO mandate by *the Dakar framework for action—education for all: meeting our collective commitments* in 2000, within the scope of the Millennium Development Goals (MDGs) to 2015. None of the goals of *Education for All* 2000–2015 were reached, and the developments in education continue to be unequal and uneven (Benavot et al. 2015). In that context, Rosa María Torres' qualitative analysis of the first decade of the EFA points to a devaluation of the key concepts. This "shrinking" occurred in the process of the "expanding of the vision" of global education. Thus, *basic education* becomes *schooling* on a primary level, leading to "frenziedly putting children in school, perpetrating the history of quantity without quality, enrollments without retention, and teaching without learning" (Torres 2020, 16). *Basic learning needs* is reduced to the *minimum needs*, *learning* is replaced by measurable *school performance*, while *lifelong learning* is reconceptualised as an end in itself. Education is seen as an isolated domain, while key elements of life outside the school—family, community, media, other socio-economic factors—are elided.[4] Education for *all* becomes education for *developing countries*, for the "poorest among the poor." Nonetheless, the recipient state is left with the responsibility and the obligation of putting education into practice. On the other hand, the rhetoric of the project places the acknowledgement of agency with the "international community." Thus, the hegemonic powers taking up the role of trustees are in "a position defined by the claim to know how others should live, to know what is best for them, to know what they need" (Li 2007, 4). Revealingly, the review of the EFA includes as an indicator *countries' adherence to the commitments*. As Torres notes, "this serves as an assessment conducted by agencies on countries, but does not allow countries to assess agencies or agencies to assess each other" (Torres 2020, 19). It is also relevant to recognise the tendency identified by Sabelo Ndlovu-Gatsheni in the case of Zimbabwe and its elites to maintain the unreformed colonial education, "education for disempowerment," in order to reproduce the effects of cheap labour and dependency (Ndlovu-Gatsheni 2021).

EFA soon transformed from declaration into a fund, the *Global Partnership for Education* (GPE), and extended its aspirations under Sustainable Development Goals (SDGs) to 2030.[5] In 2015, the *Incheon Declaration for Education 2030*, "set out a new vision for education for the next fifteen years."[6] The 2015 *Oslo Summit on Education for Development* also resulted in the establishment of "new" international institutions, such as the high-level *Education Commission* (*Commission on the Financing of Global Education Opportunities*, co-convened by Norway as "a major global initiative engaging world leaders, policymakers, and researchers"). A "new" international mechanism, the *International Finance Facility for Education* (IFFEd), was produced to fill "the critical gap in the international financing architecture for education" (IFFEd Prospectus 2020). The *Education Workforce Initiative* was also created to strengthen, diversify, and reimagine an education workforce, as was the *DeliverEd Initiative*, aiming to strengthen the ability of governments to implement reforms (DeliverEd 2020). Both were formed under the *Education Commission*, adding to the commitment "to develop a renewed and compelling investment case and financing pathway for achieving equal educational opportunity for children and young people" (Education Commission 2019). The joint mission of the "heads of states" and "leaders" in the *Education Commission* tackles "an urgent and ever-worsening learning crisis".[7] The *Education Outcomes Fund* (EOF), presented as an "ambitious effort" by a partnership between governments, donors, investors, and implementing parties, works on improving educational outcomes by promoting results-based financing.[8]

A global fund for education in emergencies and crises, *Education Cannot Wait* (ECW), launched during the World Humanitarian Summit in 2016, operates as another collaborative mechanism for political action, planning, and emergency response.[9] ECW has a similar mandate as the *Inter-agency Network for Education in Emergencies* (INEE), founded to consolidate existing networks and efforts on the occasion of the first Global Consultation on Education in Emergencies in 2000 with UNESCO, UNICEF, UNHCR, UNDP,[10] the World Bank, bilateral donors, and more than 20 non-governmental organisations involved.[11] The latest addition to the growing list is the awareness campaign, *Save our Future*, presenting itself as "a global movement of diverse voices uniting to amplify the voices of children and young people." This campaign is supported by many of the same actors within the field, such as *Global Partnership for Education*, *Education Outcomes Fund*, *Education Cannot Wait*, the *Education Commission*, UNICEF, UNHCR, and UNESCO, among others. *Save our Future* emphasises the connection between education and the SDGs, with particular focus on the disruption to education caused by COVID-19 (Save Our Future 2020).

Nordic actors play an integral role in the global education project. Their international standing as reputable welfare states is tied to an image of the region's impeccable record of public investment in national education and progressive reform. A common theme behind the Nordic countries' educational systems is their relative independence and shared goals of democracy, equality, and tolerance, defined as specific national values and practices with the aim of providing an exemplary model internationally (Antikainen 2006). However, all Nordic countries have been affected by neoliberal policies that included budget cuts, privatisation, and commercialisation in the sector, and their educational systems have undergone substantial changes (Antikainen 2006). Indeed, research on the equality and social mobility within the Nordic countries' own educational systems over the last five decades has shown that it has served as the mechanism of reproduction of social stratification and gender inequality (Antikainen 2006). Racialisation and racism continue to be present throughout the educational encounter while remaining unacknowledged due to the systematic denial of idea of "race" (Svendsen 2014). The Nordic education model remains an idealised concept rather than a practice, but this issue does not appear to hinder its international propagation. Their self-proclaimed "champion role" as top donors to global education initiatives puts Nordic countries in the position to set standards in the field, to lead and mobilise others (GPE 2017a; UNESCO 2019).

Nordic states maintain the status of generous donors and "good" Western states, seemingly neutral and with no colonial past (Engh 2009; Öhman 2010). Importantly, these states have become proverbially known as trendsetters in the matters of international development aid (NORAD 2017c). In the period 2015–2017, almost two thirds of Norwegian international aid to education (58%) was routed through multilateral funding mechanisms, mainly UNICEF (38%), *Global Partnership for Education* (17%), and various Norwegian NGOs (25%) (NORAD 2017c). According to official Norwegian aid statistics, 28% of its education aid is channelled through NGOs, while only 8% has been allocated to the public sector of the recipient country.[12] Adding "education" to the list of causes that make Norway an "exceptional" donor has contributed to the nation-branding as idealistic and overtly generous. This, in turn, facilitated the portrayal of its national foreign policy and economic interests as promoting universal values and truths. Some of the benefits of this bargain involve: the reaffirmation of positive self-identification of the national "good-doer" identity; access to global decision-making processes and venues; and access to markets and resources through development aid channels (Makt og demokrati 2003). The aspirational national character is constructed through the mobilisation of development mythologies and its engagement with the image of the "Southern Other."[13] The education project enjoys national support and a wide consensus on the "historic role of Norway in educating the world's children" (GPE 2017a).

In Finland, the development and education interventionism and the rise to the status of a donor country can be understood as a form of a bargain to escape the periphery (Takala 2001). Education, recognised as a precondition to development, is utilised as a totalising concept to obfuscate other factors relevant in development (Takala 1998). In the case of Sweden, the education and development domain is characterised by "partial perspectives—that can be termed myths—these are often thought of as objective and are, the guiding principles for current Swedish international development assistance" (Öhman 2010, 125). In this sense, the development project is primarily a public education campaign for domestic consumption, contributing to the improvement of the national self-image and nation-branding (Öhman 2010).

Similar narratives of development and education are at work in Denmark. Knowledge based on "conservative values" is posited as objective, non-political, exclusive of racism, and unresponsive to institutional relations of power. This attitude places the national standards of education into a position of excellence and superior autonomy, while evading critical considerations, such as:

> … the ways in which Western knowledge construction and colonialist practices have violated and sought to exterminate other knowledges by insisting on the universal validity of their own provincial criteria and on the 'natural' order of global capital.
>
> (Suárez-Krabbe 2012, 36–37)

The absence of public scrutiny enables the perpetuation of the myth of Danish leadership in development aid, while the underlying principles informing it have not been adjusted since the 1960s (Jensen 2018). Connections and continuities between Denmark's recent colonial history, the subsequent "Modernisation" process in Greenland, and its "Third World" development interventions is observable. Benevolence is read into both its "modernisation" of Greenland and development aid to "underdeveloped" countries. Jensen describes an intentional discursive split between Danish colonialism and Danish development aid created and maintained for the purpose of the continuation of the politically useful development project, while the colonial continuity in staffing, practices, and business interests remains uncontroversial (Jensen 2018). Development efforts are publicly perceived as positive; if failures or "corruptions" take place, this is only after the project leaves Danish hands (Suárez-Krabbe 2012, 37). In line with the spirit of enlightened sharing, the Danish strategy unambiguously refers to areas of development investment where the country possesses "special knowledge, resources and interests." According to Danish development actors, it is not the financial aspect of aid that makes the difference, but Danish knowledge and technology itself, "the nature of our engagement—experience with societal solutions, principles, values, competencies and financing" (DANIDA 2017, 12).

The gift of education

The history of education can appear as the interplay between "the mission that seeks to conserve society, its traditions, knowledge, institutions and structures, and one that seeks to transform it" (Chankseliani & Silova 2018, 9). In the case of the global education project, another layer of interaction is added by international actors, where the lines are blurred between the mission to conserve and the mission to transform, and the distinctions between exogenous and endogenous are unclear. If development intervention is understood as a paradigm that originates in the enlightened elite agency, the paternalistic attitude of trusteeship appears as *conditio sine qua non* (Cowen & Shenton 1996; Nustad 2003). The project of global education operates within the boundaries of the development doctrine, universalising Eurocentrism and the relationships of power proceeding from its epistemic hegemony.

The analysis of the case of *Education for All* as a philosophy and a practice, policy-as-discourse, demonstrates how such frameworks incorporate the contradictory exclusionary special conditions and prerequisites, making the policy objectives perpetually elusive (Peters 2007). The repetitive representations of problematised "situations" in education worldwide and reductivity of solutions offered, lead to the creation of "virtual realities and managed truths" (Tamatea 2005). Lawrence Tamatea demonstrates the potential of *Education for All* to constrain people's ability to "construct education, self and community in other ways" (2005, 313). The power to represent or "create" reality is the key element that becomes visible in the clash between the human-centred goals of *Education for All* framework and its technocratic and inhumane strategies. The framework communicates certain "truths" and "facts," while "it discursively constructs education so as to be more amenable to inscription by neo-liberal agendas" (Tamatea 2005, 313).

An analysis of the global education industry points to the visible entanglement between for-profit and non-profit actors, the public and the corporate, international bodies and national institutions (Verger et al. 2016). These operate within the conceptual framework of neoliberal public-private partnership or "creative capitalism" (Olmedo 2016). The neoliberal paradigm commodifies education as the means for achieving comparative advantage in the global market, while there is a tendency to maintain a hierarchic order of schools as "positional goods," an aspiration towards higher social status (Verger et al. 2016, 8). The conceptual framework of the global education project remains within the neoliberal confines. Some of the global education documentation aggressively promotes the competitive and "value for money" driven logic of capitalism, combining the tone of telemarketing and the language of fictitious capital investment with the rhetoric of education.[14] The technocratic solutions to the perceived education problems reaffirm the "expert" paradigm. The practice of "rendering technical" constitutes the boundary between trustees and those who are subject to education intervention (Li 2007, 7). The transactional rationales for the global education project correspond with Hlabangane's observation that Western knowledge follows the logic of maximum accumulation at whatever cost:

> Characterised by a mathematical, logocentric understanding of the world that informs the bullying tendencies mentioned above, presupposes knowledge that is universal and a knower who is not situated. Related to this is the way in which, by systematically hiding its locus of enunciation, it accords itself a 'god-eye' view of the world.
>
> (Hlabangane 2021, 167)

The education project and its planned outcomes rest on the implicit presupposition that the world economy is to remain in its current capitalist state and shape indefinitely.[15] Additionally, there is an assumption, both implicit and explicit, that future disorder, poverty, conflicts, and disasters are to occur in the South. These are attributed to population growth, health crises, migrations, and require corrective or pre-emptive measures in the form of educational development interventions. Suárez-Krabbe finds that the document *Building Stronger Universities in Developing Countries* "coherently applies notion of the backwardness of the South and the advancement of the North, thereby re-visiting the white man's burden," by proposing transfer of Northern-prescribed knowledge to the universities in the South (2012: 38). This is consistent with the Danish development strategy that proposes a number of areas of self-proclaimed expertise to be shared with the "partners" (DANIDA 2017: 12). The strategy lists education aid portfolio under Strategic Aim 1 "Security and Development—Peace, Stability and Protection," as means "to countering refugee pressures on Europe's borders" (DANIDA 2017, 5). The project's preoccupation with vocational training follows the development assumption of inadequacy. Such eagerness to provide the children of the majority world with "appropriate," "adequate," market- and employment-oriented education is based on the same assumption (Åkesson 2004).

This course of thought is reminiscent of colonial education, offering prescribed content and skills to enable unhampered governance and administration of the colony and colonial resources, concerned solely with the functionality of the colonial subjects as "end product" (Niedrig and Ydesen

2011). For example, the *White Paper 25 Education for Development (2013–2014)*, issued by the Norwegian government, puts particular emphasis on vocational education and training as priorities in international educational efforts, a solution to place "the current large youth cohort in developing countries" in the labour market (UD 2014). The project's general tone of urgency in narrating the downfall or lack of coping mechanisms of educational systems—urgency extending only to the majority world—carries the potential of creating narrative rupture. As April Biccum explains: "Not only is development not distinct from colonialism, it bears significant threads of continuity which are masked by a complex shift in vocabulary and the persistent narrative of historical rupture" (Biccum 2009). Narrative rupture works as a performative defining point which can shape our knowledge and understanding of the contemporary world (Wilson 2012). The proclamation by UNESCO highlights the sense of rupture and urgency (emphasis added in bold):

> What is **unprecedented** is **the scale at which** education systems are failing – not only to impart the basics, but **to keep pace** with **rapid changes** in global economic, socio-political and natural environments and equip students with the skills, knowledge and attitudes **to meet the challenges** they pose'.
>
> (UNESCO 2013)

The projected state of urgency provides a seemingly unquestionable rationale for the quick and efficient "placement" of the subjects of education, thus naturalising the hierarchy in the "close cooperation between education authorities and industry with regard to vocational education and training, essential to ensure that the training offered is relevant to the labor market and maintains the quality that is necessary" (UD 2014). The close cooperation becomes entangled, making it increasingly difficult to distinguish between actors, their roles, and their interests, including those of the donors. An apt example is the Norwegian *Oil for Development* programme that assists resource-rich, but "underdeveloped" countries in fighting the "resource curse" by developing their capacities, including vocational and professional training, to address a deficit of skills needed for the smooth running of the petroleum industry, directly linked to the core of Norwegian economic interests (NORAD 2017b, 2017a).

"Technological thinking" describes a rhetoric of modernity that presents technology as a universal solution for the world's problems (Mignolo 2011, 15). Educational advances are presumed to be achieved through technological improvements, with the technological packaging of options-without-alternatives.[16] Innovation, digital classrooms, app-classrooms, and e-learning are the new buzzwords (The Oslo Declaration 2015, 2; UD 2014, 30–33). Similarly, UNESCO ascribes a beneficial role to mobile technologies in the developing world as "a classroom and home learning resource" (UNESCO 2013, 10). The continued pursuit of solutions via technological thinking raises questions about the objective of the intervention and the feasibility of this strategy in the majority world. The predominant majority of the world's children have no access to modern technologies, or even the infrastructure to use them (Mignolo 2011, 15). The "feel-good" effect of the Northern intervention creates the illogical, yet conscience-soothing, notion that the benefits of technological progress can be shared to ease the suffering in the majority world. Nonetheless, innovation—while contested—is not problematic in itself.[17] Educational technologies have a decolonising potential if they offer exposure to non-hegemonic traditions of knowledge. This process, leading to an education as a "complicated conversation," needs to be informed by the lived experiences of students (Waghid 2019).

It is often assumed that the technocratic nature of international bodies makes racism or racialised representations and practices less feasible—or even impossible—by virtue of technology and detached administration. This is reinforced though the "white gaze"[18] of "experts," "committees," and "working groups." The white gaze, normalising Eurocentrism, makes European documentation and processing of "data from the field" the standard for the validation of Southern realities.[19] As Hlabangane points out, internationally recognised principles that "inform and govern ethical conduct for research on human subjects, are one such mechanism in which the Western experience and worldview are elevated to a universal status" (Hlabangane 2020, 4). In its recommendations to solve the learning crisis, the World Bank advises "to take learning seriously, start by measuring it." The Word Bank report conjures the perfidious nature of hidden ignorance amongst our children: "Without objective information on learning, parents may be unaware of the poor quality of education" (2018, 91). This information will be provided by external expert actors as part of the perpetual urgency, the same actors providing the tools, criteria, and frameworks for interpretation in accordance with their agenda (Montoya, Beeharry and Woolf 2019). Measurement tools produced by international entities such as the World Bank are purported to address what is presented as otherwise inevitable doom: "The realization that learning outcomes are poor may come only when children face poor labor market prospects, but by then it is too late" (World Bank 2018, 91). Such damage-centred research is socially and historically situated, focusing on what is deficient to document and explain failures and losses (Tuck 2009). The context, elided in the process, leads to the pathologising of the community defined by victimisation: "Without the context of racism and colonization, all we're left with is the damage, and this makes our stories vulnerable to pathologizing analyses" (Tuck 2009, 415).

The notion of "being left behind"—racing against the clock, rushing along a linear path of progress—the imposition of constant movement and time constraints, points in the direction of "competitive ways of being, doing and knowing" imposed by coloniality (Patel 2016). Continuous pressure to keep pace with newly launched standards and frameworks is placed on the recipient state, which is expected to maintain its ability to intervene in the form of the "restructuring" of the national institutions and programmatic reforms. Imported knowledge does not function in "hybrid" convergence with previous educational traditions as imagined. The abovementioned "restructurings" and reforms lead to a significant loss of institutional memory and domestic practices, thus enabling a further homogenising effect of the intervention (Antonić 2011).

Structural adjustment programmes,[20] combined with the unilaterally approved strategies of the international donors, have produced outcomes of dependency and market-driven educational systems, enhancing the inequalities and marginalisation (Babaci-Wilhite 2015, 34). Development aid comes with strings attached, and in the case of education it is "ideas and values, advice and directives on how education systems ought to be managed and targeted … not so different from the way colonizers organized education" (Babaci-Wilhite 2015, 35). Encompassing the circle of structural adjustment reform, the *Education Commission Background Report* states the increasing education expenditures are not likely to produce better outcomes if the education system is weak (Newman, King and Abdul-Hamid 2016, 20). A more recent policy paper clearly places the responsibility for failure in the Global South: "Low-performing education systems are failing more than half of the world's children" (DeliverEd 2020, 3).

Common themes that are apparent in the discourse on global education reveal paternalism and double standards.[21] These include: the insignificance of education in the Global South ("education must be put on at the top of the agenda"); the low quality of delivery ("millions of children are out of school, or in school but not learning"); education as a prerequisite to development ("quality education is essential for eradicating poverty"); "transition from school to employment" needs to be managed on a global level, as a precondition to stability and economic growth; "education helps protect girls from abuse and enables women to contribute fully to society and to economic growth" (NORAD 2017c). The *loci* of intervention, the countries where education is to be enhanced, are inferred to suffer from a chronic lack of appreciation for education. Education in the formal sense is either so neglected that it does not exist for millions of children, or it is as good as non-existent. By extension, it implies that children might be better off without it (The World Bank 2019).

Conclusions: "the learning crisis," the "girl child" and the threat from the South

Education remains a contested site of Northern intervention, opening spaces for neoliberal transformation and (neo)colonial exploitation. Despite the prevalence and the enormity of the global education intervention, a multiplicity of alternatives can be envisaged, rooted in "the responsibility to re-imagine a different world, governed by ethics that recognize the different worlds in one" (Hlabangane 2020, 10). The contradictions around the project of education in the South reveal a space of struggle.

International entities concerned with education express an attitude of urgency and emergency with their calls to awareness, campaigns, programmes, and strategies. This perpetual announcement of an "unprecedented" educational collapse in the majority world marks a new starting point for the consideration of the history of education in the South. The "learning crisis," a disaster in waiting, is "worse than previously estimated," and is threatening to "undermine the very fabric of our economies and societies" (The Education Commission 2016, 29). The learning crisis is "invisible" and "hiding," but "persistent," and only the use of "learning metrics" can assist in tackling it (The World Bank 2018). It will otherwise turn into a "learning disaster" (GPE 2021). With the COVID-19 pandemic, the learning crisis has "exponentially magnified" and turned into "generational catastrophe" (IFFEd Prospectus 2020), constituting "the greatest education emergency of our times" (Save Our Future 2020). The notions of the "learning crisis" and "education under threat" follow the logic of race-thinking and remain within the colonial archive, while implying the imminent collapse of Southern education. The education of the South is framed as a problem to be solved within the domain of Northern charitable agencies, where urgency does not permit time and space to challenge or resist the global education paradigm nor for the consideration of the legitimacy of other knowledges or educational projects. The notion of the "learning crisis," which involves the idea that the majority world education cannot be sustained without enlightened assistance, careful guidance, and nurturing, appears as an extension of the discourse of "dark vanishings," a colonial trope of inevitable extinction (Brantlinger 2003). The colonial-era discourse simultaneously aroused celebratory and mournful sentiments, asserting the necessity of decline and the denial of Indigenous futures. The discourse extended to the emerging humanitarianisms of colonialism that possessed the agency to temporarily ameliorate, but not rescue. Poverty, conflict, and migration are defined as intrinsic to the South, calling for education as a channel for the dissemination of technological thinking and a means for securitising the Southern Other. The notion of a universal "learning crisis" removes the perception of the educational project from the context, while relegating the responsibility to the recipients de-centres a wide range of economically restrictive realities of the "target" countries. This naturalises scarcity, while attributing failure to a lack of will and vision by the recipient states.

Another pivotal underlying assumption of the global education project is that the societies, cultures, and circumstances of the recipient countries uniformly produce patriarchy and sexism. Educational intervention is justified, implicitly or explicitly, by the supposedly unfulfilled lives of girls, in the tradition of "saving brown women from brown men" (Spivak 2013, 93). Thus, the essentialised "Third World Woman" reappears in the context of the global education project (Mohanty 1984). This time, personified in the image of a girl child, she appears as a "particularly privileged signifier, as object and mediator … and favored agent-as-instrument of the transnational capital's globalizing reach" (Spivak 1999, 200). The photobook illustrating the successes of the GPE's *Through the Lens of Education* reveals the focus on the racialised girl child to the extent that only 27% of the photos display images of boys (GPE 2017b). Moreover, the GPE 2020 Results Report uses only photos of girls and women as examples of GPE's geographically diverse educational activity, mostly depicting African girls (75%)[22] (GPE 2020). These portrayals are indicative of a discursive strategy of feminisation and infantilisation, the colonial representation for the South as a space inhabited by needy but deserving women and children, especially girls (Dogra 2012, 39).

The promise of education for women as a tool of economic growth professes to offer escape from imminent abuse, however it is conditioned by acquiescence to the designated role in the global neoliberal capitalism (Wilson 2015). While maintaining the level of "traditional" responsibility for home and family, enhanced with the additional burden of neoliberal entrepreneurship, the Third World Woman is projected as an unassuming pillar of the "progressive" world order. The "Third World Girl" is singled out in the educational process and assigned the responsibilities for the health and physical survival of her future family, but also for population control and success of the demographic policies:

> The girls' learning about physical well-being as well as literacy and numeracy, promises to transform family health and nutrition, providing protection against HIV infection, higher maternal and child life expectancy, reduced fertility rates and delayed marriage.
>
> (UNESCO 2013, 6)

Similarly, the World Bank celebrates the notion that "women with more schooling have lower fertility" (World Bank 2018, 41). The overall attention given to the sexuality of a young woman of the majority world invokes Edward Said's descriptions of the Orientalist preoccupation with "not only fecundity but sexual promise (and threat)" (Said 2003, 188). The interest in reduction of fertility rates through education in the contemporary development interventions is burdened by processes of racialisation and gendering. They rely on coercion and violence as the underside of the narratives of rights and choices (Wilson 2017).

In the particular context of the Nordic countries, this brings to the forefront the doctrine of saving the world through population control in the "Third World". It invokes the controversial fertility control programmes implemented by Norway and Sweden through their development interventions.[23] These programmes, now obscured, historically connect to the Nordic experience of national programmes of eugenics.[24] The programmes were practised as the teaching of racial hygiene and policies of demographic planning, both pro- and anti-natalist (Broberg and Roll-Hansen 2005; Engh 2009). The contemporary development strategies of Nordic countries maintain an interest in reducing population growth. They draw a connection between education and fertility control.[25] A Norwegian report on the category of "aid to so-called girl related projects" reveals the absence of plans for employment of the girls whose education is encouraged (Hatløy & Sommerfelt 2018). Reductionism is revealed by the condescending proclamation that "if we don't take other factors into consideration, the result might be that the girls don't get a job, but rather a higher bride price".[26]

Nordic countries' support and advocacy continues to be a driving force of the global education project, while the Nordic model is progressively becoming an ideology rather than a practice. A prevalent attitude of pride and public promotion of what Nordic countries see as the delivery of the gift of education often inhibits critical inquiry regarding its colonial embeddedness as a civilising enterprise. Indeed, the South continues to be utilised as the testing ground for neoliberal methodologies and approaches in education that carry the markers of colonial continuities, including coloniality's inherent dehumanisation.

Notes

1 This paper considers 'South' or 'Global South' in terms defined by Enrique Dussel, as the former colonial world that emerged in the 16th century and the lands and peoples that, although not directly colonised, bear the effects of European power (Dussel 2013). It is a "metaphor for human suffering under global capitalism" (Dussel cited in Mignolo 2002:66). It is important to note that neither 'South' nor 'North' are to be understood in a totalising manner or as homogenous geographical entities (Finck-Carrales & Suárez-Krabbe 2022).

2 Other externally driven policy reforms (such as the Bologna process) and transnational influences (such as European integrations) although relevant in their effect, such as homogenisation, transplanted standardisation, Europeanisation and normalisation of Eurocentrism, are not covered by this overview (Chankseliani & Silova 2018).

3 On private-public partnerships in global education and influence of corporate ethics on "assisted" forms of education, see Verger et al. (2016).

4 Some sources on global education present contextual knowledge useful to justify their mission: "There is a mistaken assumption that, as low middle-income countries grow their economies, governments are able to finance education systems through increased domestic revenues and thus targeted international funding is not necessary. But as countries transition to LMIC status they lose access to concessional financing at a faster pace than tax revenues increase, and governments must make difficult choices around allocating increasingly scarce funds among all sectors. Often these choices come at the expense of social sectors, where spending is first to get cut." (IFFEd Prospectus 2020, 12).

5 Supported by Norway, Denmark, Sweden and Finland (listed in order of the level of financial support provided). *Global Partnership for Education, Funding*, available at https://www.globalpartnership.org/funding

6 Incheon Declaration and Framework for Action for the implementation of Sustainable Development Goal 4: Ensure inclusive and equitable quality education and promote lifelong learning opportunities for all, available at https://unesdoc.unesco.org/ark:/48223/pf0000245656

7 The Education Commission webpage, https://educationcommission.org/

8 "Paying for outcomes creates powerful new partnerships and ways of working between donors, governments, foundations, and education providers. It changes the way they can achieve positive impact through a new model of funding and evaluating programs." Why Pay for Outcomes? EOF webpage, accessed 14.04.2021 https://www.educationoutcomesfund.org/why-pay-for-outcomes

9 The fund is hosted by UNICEF, while maintaining an independent governance structure, chaired by the UN Special Envoy for Global Education, Gordon Brown, also the Chair of the *Education Commission*. Available at https://www.educationcannotwait.org/ Accessed 12 April 2021.
10 United Nations Educational, Scientific and Cultural Organization (UNESCO), United Nations Children's Fund (UNICEF), the Office of the United Nations High Commissioner for Refugees (UNHCR), and the United Nations Development Project (UNDP).
11 The INEE is an unincorporated organisation without legal identity. Administration and fiscal sponsorship are provided by the International Rescue Committee and the Norwegian Refugee Council, which receive funds on INEE's behalf. https://inee.org/about-inee Accessed 14.04.2021
12 Available at https://norad.no/en/front/toolspublications/norwegian-aid-statistics/
13 The State Audit reports that the administration does not ensure reliable and relevant information about education results. Support to Results in Education for All Children (REACH) fund managed by the World Bank has not been sufficiently justified, reporting to the parliament has been limited and misleading, with no transparency on the use of Norwegian funds. In his response, the international development minister conveniently places the responsibility for absence of transparency in the following manner: "I would further point out that the assistance is aimed at people in complex life situations, who live in risky areas where conditions are often unpredictable." (Riksrevisjonen 2019: 25)
14 Example from the International Finance Facility for Education: "IFFEd offers great value for money. With $1 billion in contingent financing commitments (of which only $150 million is in paid-in cash) and $1 billion in grant funding, IFFEd could deliver a total of $5 billion in new education finance" (IFFEd Prospectus 2020, 4).
15 According to the Norwegian Ministry of Foreign Affairs: "Most of the jobs that will have to be created in the future will be in the private sector" (UD 2014). Similarly, Denmark's aid strategy envisages increased opportunities for its private sector through development assistance in areas such as climate, energy, water, food, and health (DANIDA 2017).
16 Some examples include *Visjon 2030*, available at https://norad.no/en/tilskudd/sok-stotte/hoyere-utdanning-og-forskning/visjon-2030/ and *Innovation Norway*, increasingly relevant in Norway's global development efforts, https://www.innovasjonnorge.no/ or *Global Digital Library*, available at https://digitallibrary.io/. "AI can help address many of humanity's challenges, including those related to education, the sciences, culture, media, access to information, gender equality and poverty alleviation." *UNESCO holds first global conference to promote a humanist Artificial Intelligence*, 25 February 2019, Paris, France, available at https://en.unesco.org/news/unesco-holds-first-global-conference-promote-humanist-artificial-intelligence
17 It is crucial not to lose sight of the concerns of the Indigenous and African scholars in regard of exposure to "digital colonialism," imposition of the digital infrastructure, and extraction of data (Nhemachena et al. 2020).
18 As expressed by Toni Morrison: "As though our lives have no meaning and no depth without the White gaze". Toni Morrison interview, *Colorlines*, https://www.colorlines.com/articles/tbt-when-toni-morrison-checked-charlie-rose-white-privilege
19 More on the role of scientific research as the instrument of colonial exploitation and the objectivity of the use of metrics in Smith (2012) and Walter and Andersen (2013).
20 Structural adjustment programmes are policies and measures resulting from World Bank and the International Monetary Fund involvement with national economies that include deregulation, privatisation, and cuts to the public expenditures, mostly education, public health, and public employment (Geo-Jaja 2001). They are also referred to as "agents of coloniality" and "economic colonialism through debt slavery" (Ndlovu-Gatsheni 2021: 39).

21 One of the emerging themes is "21st century skills," another contemporaneity-bound concept rooted in the competitive ways of being, doing, and knowing, used to describe "abilities and attributes that can be taught or learned in order to enhance ways of thinking, learning, working, and living in the world." (Save Our Future 2020).

22 Similarly, GPE's *Raise your Hand—A Case for Investment, GPE Financing 2025*, in their donor pitch present 73% of either individual or group photos of girls, 80% situated in Africa (GPE 2021). Norwegian *Rising to the Challenge* report included solely photos of girls (NORAD 2017c), while ECW special issue *The Fierce Urgency of Now!* presents only one photograph of boys in a classroom (ECW 2020). SIDA's education portfolio presents programmes with a particular girl/woman focus, such as UNESCO's comprehensive sexuality education through *Our Rights, Our Lives, Our Future* Programme, *Girls not Brides*, and Education for mothers and girls (SIDA 2021).

23 The family planning project in India supported by Norway (1970–1995) and Sweden (1968–1980) focused on sterilisation, and birth control. It was instigated by the Ford and Rockefeller foundations, the Population Council, and several UN agencies, "although several countries in the UN voted against this, alleging that India's programme was coercive. The Scandinavian countries supported the family planning programme, and they were among the largest bilateral donors" (Engh 2002: 42).

24 This early domestic expertise qualified Sweden and Norway as partners in population control efforts (Engh 2009).

25 A Danish strategy cites population growth as cause of poverty and conflict, instability and migration, connecting the reversal of the trend to gender equality in education and employment (DANIDA 2017). A Norwegian report on aid results cites benefits of educating girls in terms of family and national income and reduction in birth rates (NORAD 2017c: 17).

26 Statement by Gro Lindstad, *Forum for kvinner og utviklingsspørsmål* in Sørum (2018).

Bibliography

Åkesson, Gunilla. 2004. "Swedish Support to the Education Sector in Mozambique. A Retrospective Review" Final Report. Swedish Embassy in Maputo, Mozambique.

Antikainen, Ari. 2006. "In Search of the Nordic Model in Education." *Scandinavian Journal of Educational Research* 50(3):229–43. https://doi.org/10.1080/00313830600743258

Antonić, Slobodan. 2011. "Reforma obrazovanja u Srbiji i transnacionalne strukture." *Nova srpska politička misao* XIX (2): 5–25.

Babaci-Wilhite, Zehlia. 2015. *Language, Development Aid and Human Rights in Education - Curriculum Policies in Africa and Asia*. London: Palgrave Macmillan.

Benavot, Aaron et al. 2015. "Education for All 2000–2015: Review and Perspectives." *ZEP - Journal of International Educational Research and Development Education* 2.

Benavot, Aaron et al. 2016. "Education for All 2000–2015: The Influence of Global Interventions and Aid on EFA Achievements". In *The Handbook of Global Education Policy*, First Edition, edited by Karen Mundy, Andy Green, Bob Lingard, and Antoni Verger, 241–258. Hoboken: John Wiley & Sons, Ltd.

Biccum, April. 2009. *Global Citizenship and the Legacy of Empire: Marketing Development*. London and New York: Routledge.

Brantlinger, Patrick. 2003. *Dark Vanishings Discourse on the Extinction of Primitive Races, 1800–1930*. Ithaca: Cornell University Press.

Broberg, Gunnar & Nils Roll-Hansen. 2005. *Eugenics and the Welfare State: Sterilization Policy in Demark, Sweden, Norway, and Finland*. East Lansing: Michigan State University Press.

"Chair's Statement - the Oslo Declaration." 2015. *In Education for Development*. Oslo: Norwegian Ministry of Foreign Affairs.

Spivak, Gayatri Chakravorty. 1999. *Critique of Postcolonial Reason*. Cambridge, London: Harvard University Press.

Spivak, Gayatri Chakravorty. 2013. "Can the Subaltern Speak?" In *Colonial Discourse and Post-Colonial Theory: A Reader*, edited by Patrick Williams and Laura Chrisman, 66-111. New York: Routledge.

Chankseliani, Maia and Iveta Silova. 2018. *Comparing Post-Socialist Transformations: Purposes, Policies, and Practices in Education*. Oxford: Symposium Books.

Choudry, Aziz. 2009. "Challenging Colonial Amnesia in Global Justice Activism". In *Education, Decolonization and Development*, edited by Dip Kapoor, 95–110. Leiden: Brill.

Cowen, M. P. and R.W. Shenton. 1996. *Doctrines of Development*. London: Routledge.

DANIDA. 2017. *The World 2030 - Denmark's Strategy for Development Cooperation and Humanitarian Action*. Copenhagen: Ministry of Foreign Affairs of Denmark/ DANIDA.

DeliverEd. 2020. *The Challenge of Delivering for Learning - Policy Brief*. New York: The Education Commission.

Dogra, Nandita. 2012. *Representations of Global Poverty*. London: I.B.Tauris.

Dussel, Enrique. 1993. Eurocentrism and Modernity (Introduction to the Frankfurt Lectures). *boundary 2, 20/3, The Postmodernism Debate in Latin* America: 65–76. https://doi.org/10.2307/303341

Dussel, Enrique. 2013. Agenda for a South-South Philosophical Dialogue. *Budhi: A Journal of Ideas and Culture* 17(1): 1–27. DOI: 10.13185/BU2013.17101

ECW (Education Cannot Wait). 2018. "Strategic Plan 2018–2021."

ECW (Education Cannot Wait). 2020. "The Fierce Urgency of Now! Education in Emergency Response to COVID-19." https://www.educationcannotwait.org/downloads/reports-and-publications/

Education Commission. 2019. *Transforming the Education Workforce: Learning Teams for a Learning Generation*. New York: Education Commission.

Eide, Elisabeth & Anne Hege Simonsen. 2008. *Verden Skapes Hjemmefra*. Oslo: Unipub.

Engh, Sunniva. 2002. "Scandinavian Aid to the Indian Family Planning Programme, 1970–80." *Social Scientist* 30 (3–4). https://doi.org/10.2307/3518001

Engh, Sunniva. 2009. "The Conscience of the World? Swedish and Norwegian Provision of Development Aid." *Itinerario* 33. https://doi.org/10.1017/S0165115300003107

Finck-Carrales, Juan Carlos, and Julia Suárez-Krabbe (Eds.) (2022). *Transdisciplinary Thinking from the Global South: Whose Problems, Whose Solutions?* London: Routledge. https://doi.org/10.4324/9781003172413

Geo-Jaja, Macleans, and Garth Mangum. 2001. "Structural Adjustment as an Inadvertent Enemy of Human Development in Africa." *Journal of Black Studies* 32 (1): 30–49. https://doi.org/10.1177/002193470103200102

GPE (Global Partnership for Education). 2017a. "Norway Has a Historic Role to Play in Helping Educate the World's Children." Global Partnership for Education. 2017. https://www.globalpartnership.org/news/norway-has-historic-role-play-helping-educate-worlds-children

GPE (Global Partnership for Education) . 2017b. "Through the Lens of Education."

GPE (Global Partnership for Education). 2020. *The Global Partnership for Education's Results Report 2020*. Washington, D.C.: Global Partnership for Education.

GPE (Global Partnership for Education). 2021. "Raise Your Hand - A Case for Investment ."

Hatløy, Anne & Tone Sommerfelt. 2018. *Efforts to Ensure Girls' Rights in Norwegian Development Cooperation - An Analysis of Norwegian Policies from 2011 to 2017*. Oslo: Fafo.
Hermann, Martin Bille. 2018. "Denmark's Pledge, Danish Ministry for Foreign Affairs." In *GPE Replenishment Conference*. Dakar.
Hlabangane, Nokuthula. 2020. "When Ethics Fail: Unmasking the Duplicity of Eurocentric Universal Pretensions in the African Context." *International Journal of African Renaissance Studies - Multi-, Inter- and Transdisciplinarity*. https://doi.org/10.1080/18186874.2019.1620617
Hlabangane, Nokuthula. 2021. "The Underside of Modern Knowledge: An Epistemic Break from Western Science." In *Decolonising the Human - Reflections from Africa on Difference and Oppression*, edited by Melissa Steyn and William Mpofu, 164–86. Johannesburg: Wits University Press.
IFFEd Prospectus (International Finance Facility for Education). 2020. "A Powerful New Engine for Global Education." Available at https://educationcommission.org/wp-content/uploads/2020/09/200918-IFFEd-Prospectus2020-Final.pdf
Jensen, Lars. 2018. *Postcolonial Denmark - Nation Narration in a Crisis Ridden Europe*. London and New York: Routledge.
Li, Tania Murray. 2007. *The Will to Improve: Governmentality, Development, and the Practice of Politics*. Durham and London: Duke University Press.
Lugones, María. 2018. "On the Universality of Gender in Colonial Methodology." Manuscript. 9th Decolonial Summer School Middelburg. June 19–July 5. University College Roosevelt/University of Utrecht, The Netherlands.
Lundahl, Lisbeth. 2019. "Swedish Education Reform: High Ambitions and Troubling Results." In *The Conditions for Successful Education Reforms. Revue internationale d'éducation de Sèvres* (Online). https://doi.org/10.4000/ries.7843
Makt og demokrati. 2003. Sluttrapport fra makt- og demokratiutredningen. Utredning fra en forskergruppe oppnevnt ved kongelig Resolusjon 13. mars 1998. Arbeids- og administrasjonsdepartementet. Oslo.
Maldonado-Torres, Nelson. 2016a. *Outline of Ten Theses on Coloniality and Decoloniality*. Paris: Frantz Fanon Foundation.
Maldonado-Torres, Nelson. 2016b. "The Crisis of the University in the Context of Neoapartheid." In *Decolonizing the Westernized University*, edited by Ramon Grosfoguel et al.. Lanham, Boulder, New York: Lexington Books.
Maldonado-Torres, Nelson. 2017. "On the Coloniality of Human Rights." *Revista Crítica de Ciências Sociais* 114: 117–136. https://doi.org/10.4000/rccs.6793
Manthalu, Chikumbutso Herbert and Yusef Waghid. 2019. *Education for Decoloniality and Decolonisation in Africa*. Palgrave Macmillan.
Mignolo, Walter. 2000. *Local Histories/Global Designs - Coloniality, Subaltern Knowledges, and Border Thinking*. Princeton: Princeton University Press.
Mignolo, Walter. 2002. "The Geopolitics of Knowledge and the Colonial Difference." *The South Atlantic Quarterly* 101 (1). https://doi.org/10.1215/00382876-101-1-57
Mignolo, Walter. 2011. *The Darker Side of Western Modernity*. Durham and London: Duke University Press.
Mignolo, Walter and Catherine Walsh. 2018. *On Decoloniality: Concepts, Analytics, Praxis*. Durham and London: Duke University Press.
Mohanty, Chandra Talpade. 1984. "Under Western Eyes - Feminist Scholarship and Colonial Discourses." *Boundary* 2 12 (3/13). https://doi.org/10.2307/302821
Montoya, Silvia, Girindre Beeharry and Emily Woolf. 2019. "A Partnership for a Global Public Good: Data to Improve Learning." GPE (Global Partnership for

Education). 2019. https://www.globalpartnership.org/blog/partnership-global-public-good-data-improve-learning

Ndlovu-Gatsheni, Sabelo. 2021. "*Decolonisation of the Education and Epistemology Systems in Zimbabwe.*" News Hawks. 2021. https://thenewshawks.com/decolonisation-of-the-education-and-epistemology-systems-in-zimbabwe/

Ndlovu-Gatsheni, Sabelo and Patricia Pinky Ndlovu. 2021. "The Invention of Blackness on a World Scale." In *Decolonising the Human - Reflections from Africa on Difference and Oppressioning the Human*, edited by Melissa Steyn and William Mpofu. Johannesburg: Wits University Press.

Newman, John L., Elizabeth M. King and Husein Abdul-Hamid. 2016. "The Quality of Education Systems and Education Outcomes. Background Paper for the Report 'The Learning Generation: Investing in Education for a Changing World'." New York: The International Commission on Financing Global Education Opportunity.

Nhemachena, Artwell, Nokuthula Hlabangane and Maria Kaundjua. 2020. "Relationality or Hospitality in Indigenous Knowledge Systems? Big Data, Internet of Things and Technocolonialism in Africa." In *Decolonising Science, Technology, Engineering and Mathematics (STEM) in an Age of Technocolonialism*, edited by Nokuthula Hlabangane, Artwell Nhemachena & Joseph Z. Matowanyika. Bamenda: Langaa RPCIG.

Niedrig, Heike and Christian Ydesen. 2011. *Writing Postcolonial Histories of Intercultural Education*. Berlin: Peter Lang Verlag.

NORAD. 2017a. *2017 Resultatrapport: Kunnskap mot fattigdom kapasitetsutvikling av offentlig sektor i utviklingsland*. Oslo: Norad.

NORAD. 2017b. *Oil for Development Programme Annual Report 2016*. Oslo: Norad.

NORAD. 2017c. *Rising to the Challenge - Results of Norwegian Education Aid 2013–2016*. Oslo: Norad.

Nustad, Knut. 2003. *Gavens Makt: Norsk utviklingshjelp som formynderskap*. Oslo: Pax forlag.

Öhman, May-Britt. 2010. "'Sweden Helps': Efforts to Formulate the White Man's Burden for the Wealthy and Modern Swede." *Kult* 7, Special Issue Nordic Colonial Mind: 122–142.

Olmedo, Antonio. 2016. "Philantropic Governance - Cheritable Companies, the Commerciakisation of Education and That Thing Called 'Democracy'." In *The Global Education Industry*, edited by Anthony, Christopher Lubienski and Gita Steiner-Khamsi Verger. London and New York: Routledge.

Patel, Leigh. 2016. *Decolonizing Educational Research - from Ownership to Answerability*. New York and London: Routledge.

Peters, Susan J. 2007. "'Education for all?' A Historical Analysis of International Inclusive Education Policy and Individuals With Disabilities." *Journal of Disability Policy Studies* 18 (2). https://doi.org/10.1177/10442073070180020601

Quijano, Anibal. 2000. "Coloniality of Power, Eurocentrism, and Latin America." *Nepantla: Views from South* 1 (3).

Quijano, Anibal. 2007. "Coloniality and Modernity/Rationality." *Cultural Studies* 21 (2–3): 168–178.

Riksrevisjonen. 2019. *Riksrevisjonens Undersøkelse Av Informasjon Om Resultater Av Bistand Til Utdanning; Dokument 3:10*. Oslo: Riksrevisjonen.

Said, Edward W. 2003. *Orientalism*. London and New York: Penguin Books.

Save Our Future. 2020. "Averting an Education Catastrophe for the World's Children." https://saveourfuture.world/

SIDA (The Swedish International Development Cooperation Agency). 2021. "Education." https://www.sida.se/en/sidas-international-work/education

Smith, Linda Tuhiwai. 2012. *Decolonizing Methodologies*. London, Dunedin: Zed Books, Otago University Press.
Sørum, Benedicte. 2018. "Norway Prioritises Aid to Support Girls' Education, but Forgets the Jobs." Kjonsforskning. No. 2018. http://kjonnsforskning.no/en/2018/02/norway-prioritises-aid-support-girls-education-forgets-jobs
Suárez-Krabbe, Julia. 2012. "'Epistemic Coyotismo' and Transnational Collaboration: Decolonizing the Danish University." *Decolonizing the University: Practicing Pluriversity (Human Architecture: Journal of the Sociology of Self-Knowledge)* X(1): 31–44.
Suárez-Krabbe, Julia. 2016. *Race, Rights and Rebels -Alternatives to Human Rights and Development from the Global South*. London and New York: Rowman & Littlefield Press.
Svendsen, Stine H Bang. 2014. "Learning Racism in the Absence of 'Race'." *European Journal of Women's Studies* 21 (1): 9–24. https://doi.org/10.1177/1350506813507717
Takala, Tuomas. 1998. "Justifications for, and Priorities of, Development Assistance to Education--Finnish Development Cooperation in an International Perspective." *Scandinavian Journal of Educational Research* 42 (2): 177–92. https://doi.org/10.1080/0031383980420205
Takala, Tuomas. 2001. "Views of the 'Centre' vs Finland's Support to Prevocational and Vocational Education in Developing Countries." *Journal of Education and Work* 14 (2): 251–67. https://doi.org/10.1080/13639080120056682
Tamatea, Laurence. 2005. "The Dakar Framework: Constructing and Deconstructing the Global Neo-Liberal Matrix." *Globalisation Societies and Education*. https://doi.org/10.1080/14767720500166993
The Education Commission. 2016. "*The Learning Generation - Investing in Education for a Changing World*." New York: The International Commission on Financing Global Education Opportunity.
The Oslo Declaration. 2015. Oslo: Ministry of Foreign Affairs.
The World Bank. 2018. *Learning to Realize Education's Promise*. Washington, DC: The World Bank.
The World Bank. 2019. "The Education Crisis: Being in School Is Not the Same as Learning." 2019. www.worldbank.org/en/news/immersive-story/2019/01/22/pass-or-fail-how-can-the-world-do-its-homework
Torres, Rosa María. 2020. *One Decade of Education for All: The Challenge Ahead*. Buenos Aires: International Institute of Educational Planning.
Tuck, Eve. 2009. "Suspending Damage: A Letter to Communities." *Harvard Educational Review* 79 (3): 409–27.
Tumushabe, G. & Makaaru, J.A. 2013. "Investing in Our Nation's Children: Reforming Uganda's Education System for Equity, Quality, Excellence and National Development, *ACODE Policy Briefing Paper Series*, No. 27."
UD. 2014. *Meld. St. 25 (2013–2014) Melding til Stortinget - utdanning for utvikling*. Oslo: Det kongelige utenriksdepartement.
UNESCO. 2013. "The Global Learning Crisis: Why Every Child Deserves a Quality Education." Paris: United Nations Educational, Scientific and Cultural Organization
UNESCO. 2019. "Sweden Scales up Its Flexible Funding to UNESCO's Education Programmes." 2019. https://en.unesco.org/news/sweden-scales-its-flexible-funding-unescos-education-programmes
Verger, Anthony; Christopher Lubienski and Gita Steiner-Khamsi. 2016. *The Global Education Industry*. New York: Routledge.

Waghid, Faiq. 2019. "Towards Decolonisation Within University Education: On the Innovative Application of Educational Technology." In *Education for Decoloniality and Decolonisation in Africa*, edited by Chikumbutso Herbert Manthalu and Yusef Waghid. Cham: Palgrave Macmillan.

Walter, Maggie and Chris Andersen. 2013. *Indigenous Statistics A Quantitative Research Methodology*. Routledge.

Wilson, Kalpana. 2012. *Race, Racism and Development: Inerrogating History, Discourse and Practise*. London and New York: Zed Books.

Wilson, Kalpana. 2015. "Towards a Radical Re-Appropriations: Gender, Deveopment and Neoliberal Feminism." *Development and Change* 46 (4): 803–32. https://doi.org/10.1111/dech.12176

Wilson, Kalpana. 2017. "Re-Centring 'Race'in Development: Population Policies and Global Capital Accumulation in the Era of the SDGs." *Globalizations* 14 (3): 432–49. https://doi.org/10.1080/14747731.2016.1275402

6 Swedish television reporting on Venezuela as *damnation*

Juan Velásquez Atehortúa

Introduction

During the springs of 2013 through 2019, Venezuela faced several waves of violent riots stemming from the liberal opposition, backed by the ruling governments of the North Atlantic powers, specifically those of the United States of America (US) and the European Union (EU). The riots were reported extensively by the global networks of liberal corporate media. One of the Western corporate media involved in reporting the violence was the Swedish Public Television Service (SVT), whose work must be guided by Swedish governmental principles of reporting, including its obligations to provide factuality, independence and objectivity, gender equality, and multiple standpoints (Swedish Government 2013). While the SVT generally respects these principles, this chapter shows that they were suspended in its reporting of the Venezuelan crisis. Concretely, the chapter offers an in-depth analysis of the SVT reporting on Venezuela during the spring and summer of 2017, prior to the election of a National Constituent Assembly in the country. It uses two sets of news items from the SVT and ethnographic material collected by the author in the same locations used in the SVT's reporting. The first set consists of ten news articles published on the SVT's website at the beginning of the spring of 2017. This set is used to look into the factuality, independence, and objectivity claimed by the SVT. The second set of data from the SVT regards the documentary film *On the Verge of Ruin* (Norborg 2017), in which the SVT's reporting practised a coloniality of gender. In order to establish references on whether the SVT included multiple standpoints, a third set of data consists of 42 sessions of participatory observation conducted by the author between July 10 and August 2, 2017, in the same locations that the SVT reported from for the documentary.

The analysis in the chapter is informed by *the coloniality of being* (Maldonado-Torres 2007), the notion of *damnés* (Fanon 1967), *the coloniality of gender* (Lugones 2010), and *the pedagogy of cruelty* (Segato 2016). This analytical framework allows exploration of how the SVT injected suspiciousness, inferiority, and dispensability towards the brown *Chavista* population though the coloniality of being, making the reporting on the country functional to sustain a longstanding reality of *damnation* in Venezuela.

DOI: 10.4324/9781003293323-7

Additionally, the coloniality of gender is evident in the ways in which the SVT used colonial ideas of gender equality as a new brand that, in the reporting, translates into a pedagogy of cruelty that reduces human empathy to train the SVT audience to tolerate acts of cruelty against a substantial part of the Venezuelan population, which is produced as dispensable through the very same reporting.

Venezuela as an historical threat to empire

In a previous publication (Velásquez Atehortúa 2021), I argued that Venezuela has historically been a challenging problem for the North Atlantic empires during three main historical periods. The first period concerns the time when troops, heralded by the Creole elites in what is today's Venezuela, commanded the liberation army that ended Spanish colonial rule from the Caribbean to Bolivia. This period saw the formal abolition of slavery, the creation of a multitude of independent nations by 1826, and a vision to create an insular state to secure control over the Caribbean Sea together with Mexico, heralded by Simon Bolivar and his troops. Such control would also be a stepping stone leading the insular state's army over the Atlantic Ocean to liberate European populations from their colonial monarchies.

The second period concerns Venezuela's emergence as a key means for the exploitation of asphalt by the US at the beginning of the 1900s (Velásquez Atehortúa 2021). This raw material was in demand for furthering the urbanisation process in the US, and its extraction was controlled by US companies. When the government of Cipriano Castro tried to get a piece of the cake, the US led hostile economic and military actions against Venezuela (see also Coates 2015). The "asphalt war" drained Venezuela of the resources needed to pay European companies—mostly German consortiums—involved in building the county's railway. To collect the debts, these consortiums engaged their governments to conduct a blockade, and then an occupation of the country in 1903. This scenario set the grounds for the adoption of the Roosevelt corollary, by which the US redefined the Monroe Doctrine. This doctrine was initially adopted by the US in 1828 to help Great Colombia keep European empires out of the Americas and secure the independence of the nascent new states. The Roosevelt corollary states that to avoid conflicts, as in the case of Venezuela, the European powers should pass to the US the responsibility of collecting debts from the countries in the hemisphere. With the corollary, the US became a continental gendarme guarding the economic interests of European investors, securing Latin America and the Caribbean as an exclusive space for the consolidation of its growing global dominance.

The third scenario, making Venezuela a challenge for North Atlantic empires, occurred at the beginning of the 2000s with the Bolivarian revolution. Led by Hugo Chávez Frias from 1998 until his death in 2013, Venezuela adopted a new constitution grounded in participatory democracy to replace the elite rule imposed through US imperialism and neoliberal North Atlantic finance institutions, as well as control of the state by white colonial families.

Bolivarian Venezuela invited active and wide participation of the excluded and racialised mestizo/a, black, and indigenous populations, and the country channelled its surplus from oil towards four main areas. The first was to ground a general welfare state in order to improve the life conditions of the poor majority. Pivotal to this endeavour was the empowerment of barrio women in order to ground a new communal state firmly based upon grassroots organisations (Velásquez Atehortúa 2014). Second, *PetroCaribe* was created to counter the hegemony of US oil corporations over the supply of the Caribbean region (Muhr 2017). Third, the News Network *TeleSur* aimed at grounding a robust information economy in Latin America independent from national elites connected to US imperialism and disrupting the hegemonic discourse on Venezuela voiced in North Atlantic corporate media (Zweig 2017). Finally, after creating *PetroCaribe*, the government continued building several transnational anticolonial/anti-imperial blocks that concluded with the launch of the Community of Latin American States (CELAC). This new body was created to replace the dominance of the US and Canada as North Atlantic gendarmes in the hemisphere through the Organization of American States (OAE in English). This development led Barack Obama to declare Venezuela an "unusual and extraordinary threat to [US] national security" (Venezuelanalysis 2015), for example, to its hegemony in the Americas.

Naturalisation of war and the damnés

Obama's branding of Venezuela as a "threat" in 2015 further enabled corporate media to describe Venezuela as an infernal battlefield reluctant to join the "free world" embraced by North Atlantic neoliberalism. The discourse on Venezuela in corporate media placed the country in a corner that fits well into what Nelson Maldonado-Torres named a radicalisation and naturalisation of the *non-ethics of war*, meaning:

> a sort of exception to the ethics that regulates norms of conduct among Christian countries, to a more stable and longstanding reality of *damnation*. Damnation, life in hell, refers here to modern forms of colonialism which constitute a reality characterised by the naturalisation of war.
>
> (Maldonado-Torres 2007, 247)

As we will see throughout this chapter, a crucial issue in Venezuela is the significant naturalisation of war through reporting by North Atlantic corporate media since at least 2015, reporting anchored in what the Peruvian sociologist Aníbal Quijano (2000) denominates the coloniality of power—the way power relations have been historically transformed by a colonial pattern of global domination sustaining the modern world capitalist system. Coloniality refers here to the process of racialisation of colonial subjects and is based on "the radical questioning or permanent suspicion regarding the humanity of the self in question" (Maldonado-Torres 2007, 245). As will be shown, in the SVT's reporting on Venezuela, the dark-skinned masses supporting the Bolivarian revolution are misrepresented, put under suspicion, or hidden in almost every feature published in spring 2017. Inspired by *Les damnés de la terre* (*The Wretched of the Earth*) written by Frantz Fanon (1967), Maldonado-Torres additionally stresses that when war is in motion, the *damné* (the wretched) is an invisible person that emerges in a world marked by the coloniality of being. However, the *damné* is either hidden or excessively visible. The *damné* exists in the mode of not being there, which hints at the nearness of death, at the company of death (Maldonado-Torres 2007, 257). Maldonado-Torres further contends that

> the appearance of the *damné* [...] indicates the emergence of a world structured on the basis of the lack of recognition of the greater part of humanity as givers, which legitimises dynamics of possession rather than generous exchanges. This is in great part achieved through the idea of race, which suggests not only inferiority but also dispensability. From here, poverty and the nearness of death—in misery, lack of recognition, lynching, and imprisonment among so many other ways—characterise the situation of the *damné*.
>
> (Maldonado-Torres 2007, 259)

As will be made evident in this chapter, the *damné* has something to give SVT reporting, allowing it to represent non-liberal governments as perverse and non-democratic. In this move, the *damnés* are bereaved of what they have to offer in terms of visions and dreams. Thus, the *damnés* are labelled as dispensable in the reporting, while the role of superior and indispensable is attributed to the white colonial elites heading the opposition. This illustrates what Maldonado-Torres calls the condition of coloniality. Accordingly, *the coloniality of being* "refers to a process whereby the forgetfulness of ethics as a transcendental moment that founds subjectivity turns into the production of a world in which exceptions to ethical relationships become the norm" (Maldonado-Torres 2007, 259). In the reporting by the SVT, the constitution of an inferior *other* frames ethical exceptions to neglect her dreams in refusing to listen to her voices. Hence, this *other* turns into a *damné* in the reporting.

The coloniality of "gender" and the "pedagogy of cruelty"

Read through the approach advanced by Maldonado-Torres, SVT reporting on Venezuela relegates the people supporting the Venezuelan government as *damnés* (wretched), casting the country into a longstanding reality of "a life in hell." Now, the text will turn towards how this life in hell is operationalised through the coloniality of gender and the pedagogy of cruelty. Before doing so, it is important to note that in the Nordic countries, the decolonial/decoloniality is appropriated in a decorative way for opening academic doors, funding academic projects, and obtaining legitimacy for academic careers, as has been the case with the appropriation and whitening of the black feminist concept of intersectionality (Harris & Patton, 2019). Additionally, Tlostanova et al. (2019) stress that many scholars appropriate coloniality, treating the decolonial as interchangeable with post-colonial readings. The authors are also sceptical about whether other feminist scholars seriously engage with Maria Lugones' work on the coloniality of gender, just by naming it (Tlostanova et al. 2019, 292). The authors stress the importance of studying Sweden's "imperial difference" along two directions: on the one hand, to make visible Sweden's "early suspended local territorial expansion" towards the east in the Baltic region; on the other, to make visible the forced assimilation and continuing dispossession of the indigenous Sápmi population in the northern regions of the country. The authors maintain that decoloniality "can be useful to explain the imperial colonial configurations in Nordic local histories and contemporary situations" but underline the importance of avoiding the deletion of racism and dehumanisation in the process (Tlostanova et al. 2019, 292).

Indeed, Maria Lugones (2010) maintains that European racism framed the transformation of colonialism into global coloniality. To Lugones, the colonial difference established through the domination of whites over non-whites is the space wherein the coloniality of power is embodied: "To see the coloniality is to see the powerful reduction of human beings to animals, to inferiors by nature." This human–nonhuman dichotomy "imposes an ontology and a cosmology that [...] disallows all humanity, all possibility of understanding, all possibility of human communication, to dehumanized beings" (Lugones 2010, 751). Lugones speaks of a "light" and a "dark" side of gender and stresses that, precisely because of its imbrication with race, in coloniality the colonised (the dark side), being reduced to nonhumanity, have no gender. As animals, they can be female or male, but theirs is never (proper) gender. This is important to the analysis of the SVT's uses of ideas of gender equality as a brand taken by Sweden, along with other Nordic countries, to take a prominent place at the core of today's North Atlantic empires. Here, Sweden, as the role model of gender equality in the global arena, can be read through what Rita Segato (2016) calls the "village world" (615). According to Segato, colonial intervention in the village world "has minoritised everything regarding women" (615). This minoritisation frames the compartmentalisation of women's issues within a gendered binary in which the two spaces of social life are understood as being ontologically full and complete. In Segato's words,

> There is no encompassing of one by another: public space inhabited by men and their tasks, politics and mediations, business and war, does not encompass or subsume domestic space, inhabited by women and families and their many types of tasks and shared activities.
>
> (Segato 2016, 616)

This dual structure is captured and reformatted through the imposition of colonial gender binarism to frame the masculine domain of the public space. To overcome this binarism, Segato suggests to "reclaim and restore the ontological fullness and capacity to speak to the general interest of the women's space" (Segato 2016, 618). In dealing with this challenge, Segato makes two invitations:

> To take the woman question out of the ghetto and understand it as the basis and pedagogy for all other forms of power and subordination, be they racial, imperial, colonial, class-based, region and centre-periphery derived, or Eurocentric versus the rest of the world [...] and to further/ understand current forms of misogynist cruelty, we will understand what is happening not only to women and those doomed feminine, dissident, and other by patriarchy but also to society as a whole.
>
> (Segato 2016, 620)

Indeed, by applying this theoretical lens, Segato's work has made visible how conquestiality "becomes instrumental to reduce human empathy and train people to tolerate and perform acts of cruelty" (Segato 2016, 622). Segato shares Lugones' invitation to go beyond the inseparability of racialisation and capitalist exploitation to study the processes of affective reduction of people and the attempts to turn the colonised into less than human beings (Lugones 2010, 745). In this line, looking at the coloniality of gender by discerning how the pedagogy of cruelty is performed towards women, for example in news reporting, can help outline the breeding ground of the colonial psychopathic personality structure, which resonates with an abrupt and functional reduction of empathy today (Segato 2016, 623).

> In this era, suffering and aggression imposed on women's bodies as well as the spectacularization, banalisation and naturalisation of such violence, measures the decay of empathy and stands arguably as functional and instrumental to the epochal mode of exploitation.
>
> (Segato 2016, 624)

In what follows, the chapter goes into the SVT reporting, showing how suspiciousness is ascribed the dispensable other—the *damné* supporting the brown socialist government. The reporting additionally performs a form of the coloniality of gender by hiding the role of the North Atlantic powers in putting Venezuela "in hell" through an economic blockade. By focusing on the women who protest against the government, the reporting also turns into a "pedagogy of cruelty" that justifies the use of violence to overthrow the Venezuelan regime.

"On the Verge of Ruin"

On the Verge of Ruin, (Norborg 2017) is an SVT documentary made by foreign reporter Bengt Norborg. The documentary frames four stories on the situation endured by women and the urban poor in Venezuela during the spring of 2017. Three of these stories take place in or close to Chacao, the wealthiest municipality in the country located in the east of the metropolitan region of Caracas and the main political stronghold of the right-wing opposition to the government. The remaining story takes place in the *23 de Enero* district, known as one of the most Chavist barrios, in central Caracas. The first story of the documentary starts following a demonstration chanting *No hay azucar, no hay harina, en Miraflores lo que hay es cocaina* ("There is no sugar or mill, in Miraflores [the presidential palace] what you find is cocaine,"—my translation if no other stated). Norborg follows the white wealthy activists through the motorway that traverses Chacao and participates actively in breathing tear gas when the demonstration is dissipated by the Bolivarian police (Norborg 2017, min 1:47). This ingress on the film functions to describe the country as ruled with brutal police repression. In the second story, the documentary moves to show how sick women lack the most basic medication due to socialist policies. In the third story, the documentary attempts to inject "different standpoints" (with regard to SVT obligations) by interviewing a revolutionary barrio woman in the Chavista district *23 de Enero*, where the people's power appears to have replaced corrupt socialist politicians. In the fourth story, the documentary ends with a section where homeless children and teenagers hang out on the wealthiest streets of Chacao, where they survive by eating garbage from the exclusive restaurants (incidentally visited by the SVT team).

There are three kinds of *damnés* in the documentary. One kind are the children and teenagers who close the documentary. Norborg is kind to interview them to show the misery of their situation in a country which settler politicians otherwise presume to be the richest for having the largest oil reserves on earth. There are ethical and factual concerns here that would demand another chapter to explore, for example, in exposing the children—and the employees of the garbage collection company—who are eating from the garbage cans in the footage. The government was working to provide food to the population to deal with the economic blockade imposed by the North Atlantic powers. Many food businesses were inflating prices exorbitantly while also calling for cuts to the food supply to pressure protests against the government. These facts were wilfully avoided in the SVT reporting, which instead placed on the government all responsibility for poor access to food in the country. In light of Segato's approach to the pedagogy of cruelty, by hiding the role of North Atlantic sanctions, the SVT was also hiding the role of the colonial patriarchal lordship in this drama, all in all, to concentrate solely on the domestic space of women at the very ground level handling the consequences of the international blockade imposed on Venezuela.

The other *damnés* in the documentary are two kinds of barrio women. It is in line with a coloniality of gender that these barrio women are at the core of *On the Verge of Ruin*. To elucidate this coloniality, we can concentrate on the first kind of *damné*. Here the documentary follows a female activist, Vanesa Furtado, a brown leukaemic barrio woman who takes care of Lucy, her cancer-sick mother. The spectacularisation of this odd and powerful example presumably describes the life of barrio women in the country. By this approach, it could be said that the documentary tries to achieve "ontological fullness" (Segato, 2016) on the binary divide between a public space, dominated by men, and a private space of caring, dominated by women. Both Vanesa and Lucy live in a working-class barrio and try to adapt to their economic limitations and their heavily reduced access to medicine. By following Vanesa in the documentary, the audience can see how the empty groceries and pharmacies, reported in Venezuela in the North Atlantic corporate news, are delinked from the longstanding reality of damnation imposed through the imperial blockade headed by the US and other North Atlantic countries. Such a purview also avoids addressing other relations of imperial colonial sabotage with profound impacts on the everyday life of barrio women. In following her decision to support the regime change sparked by the white Euro-descendant opposition, the documentary attempts to guide the audience to adopt solidarity with her as an insurrected *damné* protesting against the socialist government.

> Vanesa: Since this dictatorial regime came to power, because this is a regime that stole your dreams, I cannot anymore. Here are we, in the struggle, and if I have to die in a march, I will do it. Since I am that sick, I am a dead woman in life.
>
> (Norborg 2017, min 3:58)

Vanesa could well fit the description of the *damné*, which "hints at the nearness of death" (Maldonado-Torres 2007, 257). She could not connect her agitation to the colonial blockade hidden in the SVT reporting. However, her subjectivity embodies authenticity when pursuing its utmost possibilities through the anti-government riots. This authenticity can only be achieved by resoluteness from an encounter with death (Maldonado-Torres 2007, 250). Bengt Norborg uses Vanesa as a bridge to continue further explaining the context:

> Oil became like a curse because you never learned to save, and you never learned to manufacture anything. You imported most and subsidised basic goods. So when oil prices fell, it went no longer. Former president Chávez and his successor Maduro are responsible for the food queues, the foreign debt, and the hopelessness of hyperinflationary Venezuela. So Chávez and Maduro's socialist party have also lost backing among old supporters, in widely known left-wing enclaves like the *23 de Enero* district.
>
> (Norborg 2017, min 12:40)

Norborg's talk is performed as a personal reflection. It frames a kind of voice from the white Westernised male *God trick* (Haraway, 1988) describing from outer space the inferiority of Venezuela as a nation which "cannot save, learn, or manufacture anything." Norborg's accusation conflates here modernity with coloniality as explained by Maldonado-Torres (2007). To substantiate this coloniality of gender further, the reporter involves a revolutionary barrio woman from the *23 de Enero* district, in a way that actualises Gayatri Spivak's parole of the white man (reporter) "saving brown women from brown men" (Spivak, 1988, 296). The SVT reporter uses the women interlocutors to show a wider suspicion with all politicians in the country: "Here, neither the government nor the opposition has any major support. Here, *you* just want a change. Zuleika Matamoros is the neighbourhood's most famous leftist activist" (Norborg 2017, min 13:51).

> *El 23 de Enero* district was always characterised as a very combative neighbourhood. I believe that the political map of it, as in the country, has been transformed. And today, the support to Maduro is not the same as the one that was with Chávez. That is, it is not even comparable. Maduro and people tell you, 'economic war'. Which economic war? Then nobody believes him. And in reality, there is no economic war. But a deep economic crisis because of an embezzlement of the nation in which high government officials and senior businesspeople have the hand tucked. And those who are paying for all that are we.
>
> (Norborg 2017, min. 14:58)

Norborg voices Zuleika's suspicion of the economic war initiated by Barack Obama in 2015 and escalated during the administration of Donald Trump. By considering the economic war as nonsense, Zuleika wants to detach herself from both the traditional oligarchs and the socialist government's corrupt class. The interview continues by making the *barrio space* of Zuleika suspicious when remarking that the *23 de Enero* district is a place "that seems more Cuban than Cuba" due to the prolific presence of murals dedicated to *El Ché* and Fidel Castro. Additionally, Norborg's description of *23 de Enero* adopts a corporate liberal description of it as an enclave of criminals and drug dealers (Norborg 2017, min 15:59), a tactic commonly used by the North Atlantic corporate press to racialise Chavista districts as populated by "mobs, thugs, and paramilitary hordes" (McLeod 2019, 62). Norborg closes this interview by stressing that Zuleika had lived in *23 de Enero* all her life; that people there were waiting for an alternative both to President Maduro, on the left, and the violent right-wing opposition, on the right. But then, Zuleika tells Norborg of her position as a third political option:

> A *third actor* must be displayed organically. It is part of the revolutionary process, but it is not with the PSUV [the United Socialist Party of Venezuela]. It does not support Maduro, and it feels that Maduro even betrayed Chávez's legacy. I think the current situation can go out of hand to both elites, say the MUD [the democratic united table] and the PSUV, in the violent escalation that we are already beginning to see.
>
> (Norborg 2017, min 17:34)

Zuleika places herself as representing a third political option able to criticise both the neoliberal right-wing opposition and the left-wing socialist government. Norborg finds himself partisan on the side of the right-wing elites criticised by Zuleika. Zuleika does not support Maduro, and her solution suggests a far-leftist third actor that she attempts to represent. Despite this, Norborg—and the SVT in the weeks to come—report only from Vanesa's point of view while remaining silent on the perspective of a *third actor* like Zuleika. This means that Zuleika's story functions only to naturalise the state of war against Venezuela declared by US foreign policy.

The "second women"

With the polarisation created in between political standpoints of Zuleika and Vanesa against the government, the SVT's reporting completely ignored women supporting the revolutionary process and the government. Vanesa became the first and privileged voice and Zuleika the instrumentalised third actor; supporters of the government and the Bolivarian revolution became non-existent "second women" who remained invisible in the reporting despite being the numerical majority. For reporters committed to the principle of including multiple voices, it would not have been very difficult to meet these "second women" heading *colectivos* behind communal councils and grassroots communes in *el 23 de Enero*. These are some of the "invited forms of insurgent citizenship" (Holston 2009; Velásquez Atehortúa 2014, 2017) sustaining the Bolivarian revolution from below (Lalander & Velásquez Atehortúa, 2013). I was invited to visit the *23 de Enero* district by one local activist from the local community radio. He was close to grassroots organisations like *Alexis Vive*, *Comuna el Panal 2021*, and *la Fundación Tres Raices*. In these organisations, women kept the revolution in motion by organising the daily production and distribution of bread in the barrio, a weekly market promoting the distribution of goods from rural areas, and locally fabricated hygiene products. They were also conducting different urban agricultural activities, such as transforming dumps into fields to grow corn and bananas and building greenhouses.

I interviewed *Marilin León* at one of the greenhouse facility and cultivation fields. She stressed how her organisation had even started an agro-tourism school project involving 900 to 1200 children from 6 to 12 years of age participating in the cultivation of tomatoes and peppers during school vacation (Femsusdev 2017, min 28:18–30:04).[1] Regarding the invisibility of their work in the corporate media, Marilin stated the following:

> So good that you can be here, to collect all these experiences, and that you can show them to your community. We hope you can promote them, to show that now in Venezuela is when there is *Chavismo* for a long time ahead.
>
> (Femsusdev 2017, min 02:24–02:38)

By interpellating my work, Marilin stressed the necessity of contesting the way in which North Atlantic corporate media, for example the SVT, make their lives, struggles, and impressive achievements inferior and dispensable in their reporting. She stressed that, in contrast to manufactured perceptions of declining support for the government and the Bolivarian revolution—as voiced in *On the Verge of Ruin* through Zuleika and Vanesa—their support for the Bolivarian process was considerable and highly visible throughout the landscape of the *23 de Enero* district. By excluding them from *On the Verge of Ruin*, the SVT negated the materiality of their work as well as the time and emotion they invested in the Bolivarian revolution. In the coming section, the chapter goes deeper into this materiality, enabling the "second women" to talk further on their struggles and achievements.

Until now the focus has been on the *23 de Enero district*, which is located on the west side of the metropolitan area of Caracas. From now on we turn to the *22 de Enero* community, located in Chacao on the East side of metropolitan Caracas.

22 de Enero and the SVT production of non-existence of the "second women"

I have travelled to Venezuela since 2010 to document the whereabouts of the Bolivarian Revolution from the perspective of barrio women in a socialist community named *22 de Enero*. This is a grassroots commune headed by landless barrio women, and on this day, January 22, 2011, *22 de Enero* seized seven underused land lots following the call made by the national government to appropriate land to build housing for the poor. In the municipality of Chacao, the members of the *22 de Enero* community lived in seven extremely overcrowded slums established among colonial villas along the main water streams traversing the municipality. Most of the inhabitants were working-class black/brown and mestizos working in the exclusive homes and exclusive hotels of the traditional white elites as janitors, gardeners, and domestic servants. Their massive seizing of land in 2011 was an example of the kind of insurgent citizenship and urbanism (Holston 2009) that took place in urban Venezuela during the time of the Bolivarian revolution (Velásquez Atehortúa 2014). The fieldwork was conducted with a participatory video approach empowering the subjects in the footage, one I call video power (Velásquez Atehortúa 2015). In line with this approach, I conducted short fieldwork visits in 2010 (May), 2011 (July–August), 2012 (August–November), 2015 (July), and 2017 (July–August).

Facing a particularly violent uprising against the Bolivarian government in the spring of 2017 on May Day, the government decided to call the election of a National Constituent Assembly. I decided then to conduct a short fieldwork visit on how barrio women in Chacao municipality tackled this election. I arrived in Caracas on July 10 and conducted fieldwork until August 2. I managed to document 42 video-recorded participatory observations in total. Six of these observations were about a *simulacro electoral*—a dry run—which was an exercise on how to go to vote. Such dry runs are organised before every election in Venezuela by the National Electoral Council, CNE. This time, the dry run was held on July 16, when the opposition called for a national strike (SVT 2017a). Five of my participatory observations from this stay involved the whereabouts of people at *22 de Enero* community in Chacao.

The *22 de Enero* community was entangled with *los pobladores*, an historic urban movement that emerged in Chile and that had disseminated through Uruguay, Brazil, and Argentina since the 1970s (Velásquez 2017). In relation to the political momentum of 2017, the *22 de Enero* community was suffering the street blockades erected by the racist *guarimbas* of the opposition. The *guarimbas* were then manning violent right-wing barricades "aimed to paralize [sic] the country and precipitate a terminal crisis for the Chávez government" (Kingsbury 2017, 784). According to Kingsbury (2017, 786) "the guarimbas extend a decades-long practice of erecting policed borders at the internal—racial and class—borders of the city." The main goal of the *22 Enero* community in 2017 was then to finish building a socialist community for 235 families on land expropriated by the Chávez government in Chacao. This exposed the community even more to the public fury of the *guarimbas* of the right-wing opposition.

I was unable to follow SVT reporting during my stay in Venezuela. However, in relation to the national strike prepared by the opposition to choke the nation (SVT 2017a), the major European airlines suspended their activities to and from Venezuela three days before and three days after election day. It appears that the SVT was so affected by this measure that it was unable to place reporters in Caracas. As a consequence, the SVT had no footage of the election to show. This situation forced the reporting to be based on chronicles, generated outside the country, that were illustrated with footage collected by freelance photographers before election day. However, in its coverage of the NCA election, the SVT seemed to be reporting on-site in Caracas, and it published 20 news articles between June 28 and August 3, 2017. The most significant part of the reporting was concentrated around election day, on July 30, 2017, with ten articles. The morning after the election day, an SVT producer contacted me asking whether I would share part of my footage for news features on the election. I said yes to this invitation and selected footage—which, as noted, was conducted among the "second women" and Chavista communities—that the producer turned down after consulting with reporters. When I returned to Sweden, I revisited events looking at the SVT's reporting. I then realised that my footage among the "second women" and their communities contrasted the total silence on these women in SVT reporting. In spite of its prolific written production, when conducting the analysis for this article I noted that the reporting did not show any footage at all. On the basis of the above-mentioned events, one can conclude that the factuality of the reporting was virtually non-existent.

In what follows, I show how the condition of damnation was experienced by the Chavista population at the same locations visited by SVT reporting in Chacao. The data consists of participatory observations recorded for video power (Velásquez Atehortúa 2015) during election day on July 30 and the day after.

The experience of damnation in Chacao

At four o'clock in the morning on election day, the motorised force drove around to clean the roads of the barriers of garbage bags and large obstacles dropped by the *guarimberos*. This force consisted mostly of motorcycle drivers who use motorbikes daily for personal use, or for making a living by quickly transporting passengers for a low price through the permanent traffic jams that characterise Caracas. During the 2017 right-wing riots, these drivers were targets of hate by the opposition, as they used to be described in corporate media as *colectivos*, meaning paramilitary forces loyal to the socialist government. According to my observations, the motorised force followed the will of the poorest communities, depending on the specific political context at issue. In relation to the political momentum of 2017, they were riding around in large groups of between 30 and 100 or more to contest the street blockades erected by the racist *guarimbas* of the opposition. During the election day in Chacao, they had to remove barbed wire and large branches from ancient trees that wealthy people in middle-class areas cut down to block traffic on the main avenues. As bus transportation companies suspended their service to boycott the election, the motorised force transported people from the barrios for free to reach the polling places that for security reasons were all located at the *22 de Enero* community. As a consequence of concentrating all ballot places at a single location, by eight in the morning, the lines at the *22 de Enero* community were enormous. The crowds were a good sign of the legitimacy of the election in facing the violent political situation in Chacao.

Despite the massive participation of people at ballot places even in the stronghold of the right-wing opposition in Venezuela, the reporting at the SVT in Stockholm was disconcerting. It claimed to have another foreign reporter, Bert Sundström, and his team on-site in Caracas. But instead of showing its own footage, SVT purchased freelance footage to inform events, with a feature published at 3:08 a.m., Caracas time, about a terror attack committed during the afternoon of election day against a convoy of police officers passing Plaza Altamira in Chacao. The camera follows the convoy of Bolivarian police officers from a close angle, zooming out the focus to show the police officers ride away. When the police reach the opposite corner, an enormous explosion occurs, injuring several officers. The white men near the camera operator were so excited about the injuries caused to the police officers that they jumped ahead, applauding (SVT 2017b). SVT removed the video from its website. However, the same video and videos taken from the same angle and location are still available on North Atlantic corporate media (CNN 2017). Due to the massive participation of people in Chacao, the *guarimbas* disappeared during election day. However, after publication of the official results at around midnight, *guarimberos* set fire to a *Barrio Adentro*, a healthcare centre run by barrio women from the Chavista grassroots which were building the *22 de Enero* community. The majority of the reporters that earlier covered the terror attack against police forces in Altamira Square avoided reporting this act of terror against a strategic welfare facility. Following Maldonado-Torres (2007), one could say that the whereabouts of over 8 million people trying to exercise their democratic right to vote were totally dispensed with in SVT reporting in order to describe the whole nation as "living in hell."

Indeed, as this chapter has shown, SVT reporting on Venezuela played an active part in the biased and highly manipulative representation of events during the spring and summer of 2017. The SVT claims to conduct its reporting following obligations to, for example, factuality, independence and objectivity, gender equality, and different standpoints. The analysis has shown that the biased preferences of the reporter were put before facts, and that the SVT failed to provide references to confirm the factuality of the reporting during this period by avoiding footage. In terms of independence and objectivity, SVT did not inform the Swedish public about the role of the North Atlantic blockade when constructing a binary division of the (male) public and (female) private spheres of the reality of politics in the country. Instead of gender equality, it appears that the white male reporter was attempting "to save brown women from brown men" (Spivak 1998, 296). Finally, in terms of multiple standpoints, the reporting opted to make visible two women representing opposite political positions but both against the government and, in this move, it made dispensable the "second women" engaged in political processes connected to the Bolivarian revolution.

With regard to the SVT's obligations, in Sweden, acting as partisan for the agenda of the opposition—to the point of celebrating violence against the police and a democratically elected government—would be nearly impossible and loudly condemned. However, in Venezuela, such identification with violent white elites was proudly undertaken in SVT reporting. This partisan standpoint exemplifies a standard that Frantz Fanon (1967) defined as a divide between *welfare* for citizens in the colonial states, and *warfare* for citizens in colonised states (Fanon 1967, 29). Through such warfare, the SVT's reporting was instrumental in enacting a pedagogy of cruelty, reducing human empathy and training its audience to tolerate further acts of cruelty against the *damnés* supporting the government. These qualities make the reporting, driven to perpetuate a relationship of coloniality towards Venezuela, *damnation*.

Note

1 All interviews conducted by the author during this period have been made open access through the ethnographic archive Feminist Sustainable Development, Femsusdev, on Vimeo. www.vimeo.com/femsusdev

References

CNN. 2017. "Explosions Rock Security Forces." July 30, 2017. Video. https://edition.cnn.com/videos/world/2017/07/30/venezuela-security-force-explosion-jpm-orig-mobile.cnn

Coates, B. (2015). Securing Hegemony Through Law: Venezuela, the U.S. Asphalt Trust, and the Uses of International Law, 1904–1909. *The Journal of American History*, 102(2): 380–405.

Fanon, F. 1967. *The Wretched of the Earth*. Translation by Constance Farrington, London: Penguin Books.

Femsusdev. 2017. "Comuna El Panal 2021 & Fundación Tres Raíces." Filmed July 26, 2017. Video. https://vimeo.com/227072876

Haraway, D. 1988. "Situated Knowledges. The Science Question in Feminism and the Privilege of Partial Perspective." *Feminist Studies* 14, no. 3:575–99.

Harris, J. & Patton, L. 2019. "Un/Doing Intersectionality through Higher Education Research." *The Journal of Higher Education* 90, no. 3: 347–72.

Holston, J. 2009. "Insurgent Citizenship in an Era of Global Urban Peripheries." *City & Society* 21: 245–67.

Lalander, R., and J. Velasquez Atehortúa. 2013. "El liderazgo femenino en la radicalización de la democracia Venezolana [The Feminine Leadership in the Radicalization of the Bolivarian Venezuelan Democracy]." *Latin American Journal of Geography and Gender* 4, no. 2: 29–44.

Lugones, M. 2010. "Toward a Decolonial Feminism." *Hypatia* 25, no. 4: 742–59.

Maldonado-Torres, N. 2007. "On the Coloniality of Being." *Cultural Studies* 21, nos. 2–3: 240–70.

McLeod, A. 2019. "Chavista 'Thugs' vs. Opposition 'Civil Society': Western Media on Venezuela." *Race & Class* 60, no. 4: 46–64.

Muhr, T. 2017. "South–South Cooperation and the Geographies of Latin America–Caribbean Integration and Development: A Socio-Spatial Approach." *Antipode* 49, no. 4: 843–66.

Norborg, B. 2017. "Vid ruinens brant." SVT Repotrarna.Video. https://www.svtplay.se/video/13669176/korrespondenterna/korrespondenterna-sasong-18-avsnitt-10?info=visa

Quijano, Aníbal. 2000. Coloniality of Power and Eurocentrism in Latin America. *International Sociology* 15, no. 2: 215–232.

Segato, R. 2016. "Patriarchy from Margin to Center: Discipline, Territoriality, and Cruelty in the Apocalyptic Phase of Capital." *South Atlantic Quarterly* 115, no. 3: 615–24.

Spivak, G. 1988. "Can the Subaltern Speak?" In Cary Nelsson and Lawrence Grossberg (editors) *Marxism and the Interpretation of Culture*, 271–313. Basingstoke: Macmillan Education.

SVT. 2017a. "Omfattande protester väntar i Venezuela." April 18, 2017. https://www.svt.se/nyheter/utrikes/omfattande-protester-vantar-i-venezuela

SVT. 2017b. "Poliser i Venezuela utsatta för bombattack." July 31, 2017. https://www.svt.se/nyheter/utrikes/poliser-i-venezuela-utsatt-for-bombattack

Swedish Government. 2013. "Tillstånd för Sveriges Television AB att sända TV och sökbar text tv." https://www.mprt.se/documents/tillst%C3%A5ndsprocesser/marks%C3%A4nd%20tv%202014/tillstand%20svt.pdf

Tlostanova, M., S. Thapar-Björkertb and I. Knobblock. 2019. "Do We Need Decolonial Feminism in Sweden?" *Nordic Journal of Feminist and Gender Research* 27, no. 4: 290–95.

Velásquez Atehortúa, J. 2014. "Barrio Women's Invited and Invented Spaces Against Urban Elitisation in Chaco, Venezuela." *Antipode* 46 (June), no. 3: 835–56.

Velásquez Atehortúa, J. 2015. "Episodes of Video Power Supporting Barrio Women in Chacao, Venezuela." *Area* 47, no. 3: 327–33.

Velásquez Atehortúa, J. 2017. "Barrio Women's Gendering Practices for Sustainable Urbanism in Caracas, Venezuela." In *Women, Urbanization and Sustainability: Practices of Survival, Adaptability and Resistance*, 65–88. London: Palgrave McMillan.

Velásquez Atehortúa, J. 2021. "The Fear for Another Revolution/Colonialism – The Evolution of the Monroe Doctrine as an Instrument of Racist Domination and Hegemony in the Caribbean." In *Beyond the Market – Social Inclusion and Globalization*, 157–73. London/New York: Routledge.

Venezuelanalysis. 2015. "U.S. President Barack Obama Brands Venezuela a 'Security Threat,' Implements New Sanctions." March 9, 2015. https://venezuelanalysis.com/NoZL

Zweig, N. 2017. "Televising the Revolution as Cultural Policy: Bolivarian State Broadcasting as Nation Building." *Global Media and Communication* 13, no. 2: 181–94.

7 Creolizing subjectivities and relationalities within Roma-gadje research collaborations

Ioana Țîștea[1] *and Gabriela Băncuță*

Introduction

Historically, Roma-related research has been marked by Gypsylorism, an equivalent of Orientalism in studying Europe's internal "others" or, as Ken Lee puts it, "Whilst Orientalism is the construction of the exotic Other *outside* Europe, Gypsylorism is the construction of the exotic Other *within* Europe—Romanies are the 'Orientals within'" (Lee 2000, 132). A vast amount of Roma-related research today still reproduces Gypsylorist tropes (Matache 2016, 2017; Selling 2018). In parallel, a new critical paradigm in Romani studies has been emerging during the past couple of decades, addressing the persisting exclusion of Roma contributions from knowledge production and decision-making on and for the Roma, arguing for more critically reflexive, collaborative, and Roma-led studies, and bringing Romani studies in dialogue with critical race and whiteness, queer, post- and decolonial feminist studies (Ryder et al. 2015; Bogdan et al. 2018; Brooks, Clark, and Rostas 2021).

Researchers employing critical approaches have shown increased interest in Romanian and Bulgarian Roma migrants in Finland, yet this research remains scarce (Tervonen and Enache 2017; Keskinen et al. 2018; Himanen 2019), it presents language barriers and the need for interpreters and mediators, undermining the development of trust and collaboration with the Roma participants (Enache 2020), and very few studies are carried out or led by Roma themselves (Gheorghe and Mocanu 2021). In 2020, Ioana was invited to conduct interviews with 15 Romanian Roma women living in Helsinki, for a research project exploring intersectional discriminations experienced by Roma women in Finland, Romania, and Italy. Gabriela was one of the interviewed women. The project was Roma-led and some of the interviewers, when and where possible, were Roma (Gheorghe and Mocanu 2021). For this chapter, we wanted to go beyond interviewing methods, and argue for collaborative, co-authorial, and reciprocal forms of dialogue in qualitative research, through which we hope to challenge structural divides and power hierarchies between Roma and gadje[2] worlds. But even if we try to cross divides and hierarchies, they continue to shape and influence our interactions and conversations and cannot be done away with through our collaboration. We

DOI: 10.4324/9781003293323-8

therefore stay with the discomfort of these tensions, as productive failures from which to disrupt established ways of seeking knowledge.

In 2021, we were both offered jobs within another project offering employment opportunities to Romanian and Bulgarian Roma women living in Finland, where we worked together for one year. Roma women with little or no formal schooling or language skills—including Gabriela—were assigned to do cleaning work; Romanian and Bulgarian gadje women from privileged socio-economic positions—including Ioana—were hired as translators and mediators; and Finnish gadje women ran and sponsored the project or promoted its cleaning services to potential Finnish clients. Finnish women also constituted the majority in the decision-making board. None of the Roma women were part of the board. The project thus reproduced racialised, gendered, and classed hierarchies, with Gabriela at the bottom and Ioana in the middle of the hierarchy. Occasional tensions in claiming ownership over the project occurred. Roma women rightfully saw the existence of the project and of the other women's jobs as depending on their hard labour and wanted more participation in decision-making. Romanian gadje women made paternalistic claims to ownership based on their perceived knowledge of the Roma women's "needs" and on the perceived centrality of their translation services to the project. Finnish women claimed ownership based on financially sponsoring the project or on bringing financial resources from the clients they sought, while claiming to "empower" Roma women through low-paid precarious labour and to "help" them reach the "right" level of "development" through an ideology of assimilation into the racial capitalist order (Vergès 2021, 14).

Creolization and im/purity

The power hierarchies between us raise the question of whether Ioana is tokenizing Gabriela as her co-author, "empowering" Gabriela in Ioana's own terms. To address this issue, we work through, against, and beyond those hierarchies. We work through them by revealing how our worlds fit into the hierarchical power relations of the racial capitalist order in which we find ourselves, and how that demands both of our different roles in the various positions we undertake. We work against and beyond them through an ongoing, open-ended process of becoming through which we contest, re/negotiate, and destabilise the boundaries and hierarchies within and between each other and our worlds (Monahan 2011, 195). In other words, we creolize our subjectivities and relationalities.

Creolization emerged from the specific historical context of the Caribbean, marked by colonialism, slavery, racial classification, forced displacement, loss of social identity, and a double consciousness based on experiences of oppression and struggles for liberation (Glissant 1997; Du Bois 2005). Roma people have also been said to have developed a double consciousness and a creolized diasporic subjectivity due to experiencing historical displacement in relation to India, movements in multiple directions, and current feelings of exile in relation to countries of residence either as citizens or as migrants

(Le Bas 2010). Furthermore, Roma people have also been forced into slavery on the territory of what is now Romania from 1370 to 1856, during which they could be owned, bought, sold, donated, left as inheritance, given as treasury, and forced into various forms of coerced labour and brutal punishment (Matache and Bhabha 2021). Moreover, Roma people were also subjected to racial classification and genocide as part of eugenicist nation-building projects all over Europe (Turda 2010).

While creolization emerges from situations marked by severe inequalities and oppression, it also reveals new ways of understanding the world as relational and interdependent, marked by multiple, unexpected, transversal encounters, connections, and becomings (Glissant 1997; Gutiérrez Rodríguez 2015). Creolization is not the same as cultural or ethnic mixing, as it bypasses any racial, ethnic, and socio-cultural classifications, yet it also emerges within racialised configurations due to existing legacies of colonial practices (Gutiérrez Rodríguez 2015, 94). Creolization helps understand the ties between people and worlds that were supposed to be radically unequal and separated, by bringing them into conversations that "could not have taken place historically but that would have been and still remain generative" (Gordon and Cornell 2021, 1).

Individuals may inhabit simultaneously several distinct and separated worlds, which do not communicate or understand each other. We may disagree with how we are perceived in some worlds, though we may also internalise and animate perceptions we disagree with (Lugones 2003, 78). Social fragmentation thus prevents individuals from communicating with each other, as well as fragmenting each individual's subjectivity into parts that do not fit well together. This multiple fragmentation cuts ties within/between individuals and within/between their worlds, reproducing what María Lugones (2003) calls the *politics of purity* or what Édouard Glissant (1997) calls the *duality of self-perception*, separating individuals and worlds/communities into distinct, hierarchical, pure, homogenous categories, which are positioned as "threateningly" opaque to each other, thus making them easier to order and control.

Creolization decolonises the politics of purity. Our chapter thus tries to reject the fragmentation of our subjectivities and worlds into pure parts by working against the politics of purity, even as we find that we have them internalised. We enter this process of inter-subjective creolization from different power positions that condition the kinds of interventions that are possible for each of us. Nevertheless, as our storytelling shows, we can shape the ongoing contestation of meanings and boundaries and change how they condition our different agencies toward unforeseeable results (Glissant 1997, 34; Monahan 2011, 205).

Storytelling: methodological and ethical considerations

Storytelling is highly regarded in Romani culture. Stories transmitted via generations in various forms and channels build a "multifarious" history of the Roma that "insurrects hegemonic history" (Costache 2018, 42). The

Critical Romani Studies journal's 2021 special issue on Romani literature suggests that oral stories should be given as much significance as written ones to convey a comprehensive Romani literary canon, while arguing that literature is a very important dimension of Romani culture. The issue includes a collection of stories by Romani creative writers and storytellers, alongside its more conventional academic articles. The authors' stories, which explore the diasporic, hybrid, and multilingual characteristics of Romani literature, serve as a starting point for exploring the multidimensionality of Romani narratives (Martín Sevillano and Marafioti 2021).

For our chapter, we bring Romani storytelling epistemologies and methodologies in dialogue with María Lugones (2003) and Édouard Glissant (1997), by using storytelling as a critical tool to travel to each other's worlds, unlearn internalised perceptions based on socio-cultural classifications, and understand the inter-relational and interconnected character of our complex beings. Through dialogical self-reflection and self-aware experimentation and mutual identification, we look at ourselves in each other's mirrors and back in our own to see with each other's eyes (Lugones 2003, 84). We thus try to understand and unlearn the long-lasting legacies of slavery, eugenics, and "modernisation" practices, which position Roma women within socio-economic dependency, and teach gadje women arrogant perceptions that inferiorise Roma women (Lugones 2003, 71).

Over the course of two years, we told, read, and sent each other stories about and beyond migration, from similar yet differently inhabited local, trans/national, inter-generational, familial, social, institutional, work, and ordinary everyday contexts. We mainly worked in Ioana's studio, which was very close to the social.[3] At Gabriela's signal, Ioana would start recording some of our conversations. Gabriela also recorded stories with her phone and sent them to Ioana to be included in the chapter. Gabriela told stories about her life—from childhood to present day—spontaneously, depending on what triggered her memory at a specific moment or during the course of a given day. Emerging from emotions and memories triggered by Gabriela's stories, Ioana wrote short stories in Romanian, to which she also added some theoretical reflections. Ioana then read both her stories and Gabriela's transcribed stories to Gabriela who expressed dis/agreements and asked for changes. Ioana took notes and applied those changes. She then translated the stories from Romanian to English, attempting to be as faithful as possible to Gabriela's tone and style. Gabriela further included letters and emails she received from the Finnish Immigration Service (Migri), the police, and her lawyer during her process of applying for EU citizen residence, so that migrants facing similar situations may find sources of inspiration and strength. Ioana translated those letters/emails from Finnish to Romanian/English.

Furthermore, Ioana wrote in English the other parts of the text. Since Gabriela does not read English and only speaks a few words, Ioana summarised and translated these parts into Romanian and read them to Gabriela while explaining as best she could the various concepts and theories Gabriela was not familiar with. We also discussed how this text relates to Ioana's PhD project and went deeper into what the project is about. However, it was often difficult to negotiate the great distance between Ioana's research aims and theories and Gabriela's everyday life. Therefore, misunderstandings might have occurred, and Ioana's account might only have been imprecise and distorted. Given these ethical dilemmas, Gabriela's inclusion within academic co-authorship may still be seen as epistemic exploitation. That is because doing research *with* rather than *on* participants still requires institutional changes regarding co-research, co-writing, co-authorship, ethics, and what counts as knowledge (Sinha and Back 2014). Our text is a small contribution to wider efforts to open research towards collaborating and co-authoring with participants who are not affiliated with institutions or do not have university degrees, thus making an ethical and political statement against framing universities as the only sources of valid knowledge (Soares, Bill, and Athayde 2005; Back, Sinha, and Bryan 2012; Gay y Blasco and Hernández 2020). Yet while opening universities to diverse bodies and plural knowledges may decrease the harm they exert, incorporating alternative knowledges within academic publishing practices may also re-assert universities' hegemony. We therefore share our failures as disruptive sources of knowledge without knowing the potential results.

Stories of world-travelling

We discussed which of our individual stories to keep or leave out. We identified a few themes and combined the individual stories—together with the letters/emails from institutions—according to those themes into eight collective stories. Ioana read the eight stories to Gabriela and we both reflected on them critically. We included our mutual reflections as additional layers in the stories. The stories thus weave together multiple speaking/writing genres and multiple diffracting layers. We speak together and apart, with each other, about each other, with ourselves, and with the readers, sometimes reflecting on or critiquing the other's accounts, thus asserting our divergent agencies and "impure" states toward becoming ambiguous, unclassifiable, unmanageable (Lugones 2003, 100). The stories do not offer definite closures, but rather moments of transition to other stories or to other worlds, travelling from one time/space to another through affective connections, interferences, dis/harmonies, and transversal encounters (Glissant 1997, 58, 199). Each collective story is thus a piece in the ongoing puzzle of creolizing our subjectivities and relationalities.

Loud silence

GABRIELA (G): As Ioana and I are talking one evening in her studio over a glass of wine, I recount: "You know how the social came to be? In the early days, 10–11 years ago, we were sleeping in an abandoned building, an old train station, where they built this new library now in Helsinki,

Oodi.[4] That abandoned building didn't have doors or windows, nothing. The only thing we had was a roof above our heads so it wouldn't rain or snow on us. There was thick ice on the walls during winter. We were 50–60 people in that building at some point. I stayed there on and off for three years. When we went there in the evening, we went at 22:00–23:00, so the police wouldn't see us and chase us. We jumped the fence one by one. We had to climb, there was a big fence. And in the winter, some slipped, some fell … We'd get sick often … how many hospitals, how many treatments … When we came out from that cold from the abandoned building, we went to the train station and sat by the heaters, because that was our spot. And the security guards would come and chase us out. Good thing there was another library, where there's the market now, to sit down and get warm. But they'd chase us sometimes from there also because we were not allowed to talk or sleep there. Gadje activists and researchers became interested in us sleeping in that abandoned building and were interviewing us all the time. Even a film was made. In the end, the police found out that we were sleeping there. They came at night and chased us out, they pepper sprayed our eyes. They also sprayed that building so we won't go back in. That's how it started. We continued being on the streets, sleeping rough, giving interviews. Gadje looked at us like aliens. We were not asylum seekers, so could not sleep in a reception centre. We were EU citizens, but we were homeless. Yet we were not Finnish citizens, so could not sleep in a homeless shelter. After many struggles, the social was opened for us where we could sleep but also do much more, like talk with Finnish people and find work. But in the end gadje took all the credit and all the leadership roles. They use against us Roma that we have less education, and they silence us. All our complaints go back to those we complain about. Some of us were intimidated when we talked too much, outside of the social."

IOANA (I): You know what the gadjo leading the social told me today? That I should be more authoritative with you. That when I come to work, you must fear me. That I should not be your friend because you'll get lazy.

G: Did you talk about this with your boss?

I: No, but I told my Finnish co-worker. She said, "Jesus! He *NEVER* does that with me! I guess he does it depending on your cultural background, given you're both from the Balkans. He's an important client for our project with Roma cleaners, but that doesn't mean he can be disrespectful to you."

G: Gadje have this habit of making everything about themselves. I nod in disapproval at Ioana's story and tell her, "That's really bad." I know that's what she wants to hear. Then I start browsing a printed issue of a magazine I used to sell on the streets of Helsinki, before I started working as a cleaner. "They wrote a story about me in Finnish. From an interview I gave them many months ago. They also printed my photo, look. Tell me what it says."

I: They wrote you're homeless and begging on the streets of Helsinki to send money to your children back in Romania. They tell readers to donate money to some NGO to help vulnerable Roma.

G: Really? That's all they wrote from everything I told them? That's what they used my photo for? To make people pity me?

I: Did they tell you they were going to use your photo and story to raise money?

G: Yes, but I thought they would tell a better story. I told them so many things … You know I love to tell stories.

I: "These stories we tell each other … We could write something together. And have both our names on it." I suggest timidly. Gabriela gazes out the window in silence, as if contemplating on what I just said. "I'll do this with you on your own terms. And you're free to change your mind and quit at any time, I'll respect your decision," I continue. Gabriela remains silent, looking at her phone, scrolling up and down. I try again, "This might shake a bit how things are usually done when Roma work with gadje. Although once the story is out there, it's out of our control and people reading it may give it completely different meanings from what we intended …." Gabriela turns to me, looking deep into my eyes, as if trying to read my thoughts. Then she changes the topic of conversation.

A morning swim

IOANA: We arrive early for Gabriela's appointment with Migri. To pass the time, we take a walk by the sea, across the street from the office. It's a sunny September morning, warm enough for a light jacket, the tree leaves still green, soft sunrays dancing with their water reflections. We walk toward a wooden structure by the sea where people can wash carpets or dive and swim.

GABRIELA: A gadjo is laying against his back on a table. Amused, I ask if he's sleeping. He's soon joined by a gadji. Maybe she thought we were trying to steal him from her. They remove their clothes and dive in the water for a morning swim. She swims faster than him. But the water must be so cold! I walk to the edge and test the water with my hand. It's freezing!

I: As we admire the white heteronormative couple and the woman's strength to swim in freezing water, even faster than the man, we contemplate how maybe one day Gabriela will also have that "freedom," if only her Migri application goes through. A white feminist fantasy come true. Aspirations to equal privileges as those granted to white men by white supremacy erase gadje women's complicities with white supremacy (Vergès 2021, 12). We enter the office. All the documents for Gabriela's EU work-related registration are in order, her work contract, last three months' pay slips, and bank statements.

"It will take a few months to process the application. You will be notified if we need additional information," says the clerk.

G: Once we step outside, I can finally breathe. "I thought about it."

I: About what?

G: About writing something together. I want to leave something behind, a story in my own words. Something my children could one day read and feel proud.

The pan in the system

IOANA: My phone rings. One of Gabriela's co-workers is on the other end. "The police stopped Gabi at the airport! I was just calling to wish her a safe flight, and this shit now! They're keeping her in a room, maybe she'll miss her flight!" she says frantically. I immediately call Gabriela, but her phone is off. I try again and again. Nothing. I'm starting to panic.

GABRIELA: I'm trying to travel to Romania to see my children, whom I haven't seen in over two years. If I had my children here next to me, I'd have a different life. I'd be growing young. Like this, being far away from my children, I'm growing old … Why is he keeping me in this room, I want to ask the Finnish police, but I can only say *miksi*—why—in Finnish. Emotionless, he just says *odotta*—wait—in Finnish. A gadji walks into the room. "You're red flagged in the system. I'm here to help you," she tells me in Romanian. "Help" from gadje comes in many forms, but rarely one that's helpful.

POLICE (P): What do you do in Finland?

G: I work with a contract. You can check the letter from my employer.

P: How will you feel if we will not let you return to Finland?

G: I feel the weight of those words. I also feel sick, like I ate something bad. But I don't tell her this. Instead, I tell her "My employer needs me back to work after my trip to Romania." She calls Ioana whose number is on the letter. She needs another gadji to validate my story. Or to overturn my story and validate her gadje thoughts.

I: My phone rings again, this time a number I don't know. I hope it's Gabriela.

P: I'm calling you from Helsinki Airport. Gabriela is red flagged in our system.

I: Why is she red flagged?

P: She's a potential accomplice in stealing a pan.

G: Finnish and Romanian police working together in the big case of the missing pan. Who took the pan? Who helped to take the pan? Who will be deported for a missing pan?

P: Did you do these documents for her?

I: Yes. I represent her employer. Are you the interpreter?

P: Yes, and more.

I: Will you let her board the flight?

P: Depends on her, if she cooperates.

G: My story verifies. Free to fly. But don't get your hopes up. Freedom in gadje's terms always comes with restrictions.

P: Come back within a week to continue with the investigation. If not, we will ban you from entering Finland.

G: Food cooked in the pan on an electric stove is not even that good. The best food in the world is cooked in a cauldron on the fire.

My mother makes the best polenta in the world

GABRIELA: My mother made food in a big cauldron, delicious stews with meat and vegetables served with polenta made from corn flour. She used the vegetables she herself planted in our garden and the animals she raised in our yard. She raised pigs and chicken, and in the garden, she planted onions, potatoes, tomatoes, carrots, everything we needed. Us children helped her out with the work around the house. We were seven children, four girls and three boys. Sometimes our mother brought the cauldron in front of the house, on the street, for people to eat together, and we all sat together like that, adults and children, and ate, talked, and played … A Finnish artist who came to paint the walls of the social a few years ago turned my story into this painting (Figure 7.1). But she only painted the polenta. I laughed when I saw that. How can all of us be satisfied with just polenta? My mother also cooked meat and vegetables to go with it. But I told the artist she did a good job. Although I couldn't read then what she wrote above the painting. Only later I found out it says "my mother makes the best polenta in the world." As a child, I didn't go to school. My sisters didn't go either. My brothers went. I told my parents I wanted to go to school. They'd tell me to wait for my

Figure 7.1 My mother makes the best polenta in the world (Photo taken by Gabriela with her phone).

brothers to come home and show me their homework so I can learn from them. That made me very upset because I wanted to go to school just like them. In other Roma communities, girls go to school, but where I grew up girls didn't go to school. But I know from my brothers that it was still hard for them because all the time they heard from their gadje teachers and classmates, "You won't succeed because you're Gypsies and Gypsies are not made for school" … I started learning how to read and write a couple of years ago. One of the Romanian workers at the social, she taught a few Roma women, to "empower" us as she put it, so that if we go somewhere we can write our names and sign documents. Gadje like to use this word with us, "empower" … They want to "save" us. They think our Romani 'tradition' is different and say, "Oh no, look at these Gypsies, they're so 'backward'." But we don't need saving. We need to work together. Us Roma, we're not racist against gadje. We're happy to work together with gadje because we don't have that priority a gadji has who can enter anywhere …

IOANA: My grandmother worked on a farm, picking peaches and apples from the orchard, tilling the soil, cleaning. She lived near the farm with her husband and their five daughters. They raised their own animals, planted their own food, and baked bread in a stone oven outside in the yard. My mother said, "We lived … not better or worse than children today. The red hair bows and our mother's songs protected us from 'the evil eye'." My grandmother didn't go to school as a child. She attended three classes later in life due to a state socialist[5] policy targeting illiterate peasants. Her daughters earned degrees and had successful careers in a state socialist system that granted women from privileged ethnic majorities education and employment opportunities and leadership roles. Gender equality discourses did not extend to the private family sphere, though, so gadje women were seen as mothers of the socialist nation, assigned with reproductive roles of "proper" (white) socialist subjects (Todorova 2018, 122). My mother said, "The brutal, badly internally and externally orchestrated change of 1989 found us with a three-week-old baby. We lived the so-called 'transition' to capitalism together and tried to make our children's lives more beautiful. Raising children is difficult in any époque. You can follow all common-sense guidelines, but circumstances will still be more decisive." Gadje women also led state socialist campaigns seeking to "modernise" Roma women, like cultural eradication, forced sterilisations, and "socially useful" reforming education and labour programmes; Roma people's resistance to assimilation and the preservation of their cultures and values were framed as "backwardness" (Todorova 2018, 123). My mother tolerated my friendship with a Roma girl from school, yet she told me to never eat the food she's eating as I might get food poisoning.

G: When I cooked food at home, back when I lived with my children and husband in Romania, I always put an extra plate on the table and told him, this one's for your gadji. Once I caught him. I was with my first two children, the third one wasn't born yet. He saw us through the window approaching the bar. He quickly moved to a different table. I sat at the table with the gadji he'd been sitting with, my children next to me. I poured myself a glass of brandy from the bottle they'd been sharing. I raised my glass towards her and said cheers, looking straight into her eyes. After some hesitation, looking around as if waiting for someone to save her, she raised her glass also, with a dumb smile on her face. I took a sip from my drink. Then I walked over where my husband was sitting and poured the rest of it on his head. The gadji told me, "Sorry, I didn't know he was married and with children, and that he was a Gypsy." She probably realised that from the way I was dressed. Clothing and language, that's mainly what differentiates Roma and gadje. And sometimes skin colour. I'm a bit brunette, but there are others who are darker. Although there are also Roma people who are blonde … My first two children were born with dark skin and brown eyes, but the third one had light skin and green eyes. When my husband saw her, he said, "Now you've given me children!" For a long time, I believed light skin is beautiful and dark skin is ugly. Now I'm proud of my colour and I teach my children the same. May God also gift others with darker skin colours because there are many who try to tan but we're natural. Still, I cannot say one is more beautiful and another one is uglier, only the soul matters …

Keep walking

MIGRI (M): The Finnish Immigration Service considers deporting you and imposing an entry ban.[6]

LAWYER (L): Migri sent you a standard letter they send to all applicants whom they target with potential deportation for often unfounded reasons. Many applicants receive this letter, particularly non-EU ones.[7]

GABRIELA: My mother used to say, dear, don't argue with anyone. Even if they try to harm you. God will take care of them.[8]

M: You are given the opportunity to respond in writing, in your own words, to the following questions. How do you feel about being deported and receiving a ban on entering Finland?

L: In your case, they sent the letter due to the ongoing police investigation. This is a small misdemeanour and you are only seen as a potential accomplice. You cannot be considered an accomplice if you had no prior knowledge of the other person's intention to steal the pan. If found guilty, you will only receive a fine based on the value of the pan.

G: Don't offend anyone, dear. Don't dwell on what they say or do. Pretend you didn't notice.

M: Do you have family members, other close relationships or a job in Finland or another Schengen country? How would a deportation or re-entry ban affect these relationships or your job?

L: According to the Finnish Aliens Act, EU citizens can be deported or banned from entry if they are considered a danger to public order and security.

G: Don't hold grudges with anyone even if they hold grudges with you because not even God holds grudges.

M: You can use an assistant when sending the response. If you want to use an assistant, you need to get one yourself.

L: Yet Migri has enforced many deportations of Romanians and Bulgarians lately, usually for petty crimes. You can also be deported when you have been on social welfare for long periods of time and thus considered a burden to society.

G: Don't give too many explanations, don't excuse yourself. People will be talking about you anyway. They will make up their own story depending on what suits them.

M: You must submit your response within seven days of receiving this letter. If not, the Finnish Immigration Service will remove you from the country and impose an entry ban.

L: In response, tell Migri as little as possible, do not give them reasons to further investigate anything.

G: My mother taught me how to leave space for hello with everyone, even in the most difficult situations.

L: Say you are an EU citizen exercising your right to seek employment in another member state. You have a job in Finland, a good salary, and receive no social benefits. Mention that you have been called to the police station, but you have not committed any crimes in Finland, and are not a threat to public order and security.

G: She taught me to talk to everyone, even if there are many hateful people and maybe they talk bad about you, don't react. You watch, listen, observe, learn …

M: You will not receive a further notification.

L: Firmly and politely oppose a potential deportation, which would not be based on the law.

G: Don't react to provocations. Put them under your foot and keep walking. And take care of your children, sisters, brothers, parents, elders, your work and goals …

Able to defend myself

"Who are you?" asks the police officer at the reception desk.

IOANA: I'm Gabriela's work supervisor … and her moral support person during the interrogation.

POLICE: You cannot go inside with her.

I: "What if I have this?" I hand over the power of attorney letter signed by Gabriela.

"Alright …" answers the officer, slightly annoyed. "You can go."

After going through security check, we're greeted by another police officer: "The interpreter couldn't make it. We're trying to get one by phone."

I: I could translate.

P: We'll go with our official interpreter. Please wait in the lobby while I make some phone calls.

After more than one hour waiting, the officer comes out. "We couldn't get an interpreter by phone either. You can translate."

Gabriela: On our way to the interrogation room, I tell Ioana, 'A good spirit must be watching over us today.'

P: How do you plead?

G: Not guilty.

The officer plays for us the security camera footage.

P: Do you know the person caught on camera?

G: Yes, we know each other.

P: Did you help her take the pan?

G: "Can we just say we'll pay for the pan and they should leave us alone? I don't want to get Lili[9] into trouble." I tell Ioana.

I: But if you offer to pay for it, they'll think you helped steal it. And then they'll deport you. The footage clearly shows Lili taking the pan. The question now is whether you will also be affected by her action. Lili already has an EU residence permit, she's in a better position than you. Please, let's tell him what the lawyer advised us.

G: You and your gadje ways … Alright, go ahead …

I: I had no prior knowledge of any intention to take a pan. I paid for my product, as the footage shows.

P: Is that all you just discussed with her?

I: He's doubting us.

G: Tell him about the police making us sign something we didn't understand.

I: When the security guards stopped us at the store, and when the police took us to the station, they did not talk to us in a language we would understand. At the station, we were asked to sign a document we could not read. We asked for a translator, but did not get one.

P: The store does not want to settle. So, if the other person's story does not match yours, this will go to court. Do you agree with your written testimony being used in court?

I: "Should I answer, yes?" I ask Gabriela hesitantly.

G: Do you think that's wise? Ask him what it means.

I: What does that mean exactly?

P: If this goes to court, you don't have to go there in person and this signed statement can be used in your absence.

G: Don't give him permission to use some paper against me. I want to be able to defend myself.

Gabriela receives a letter from the police shortly after our visit at the station[10]:

> In the pre-trial investigation, Gabriela denied her guilt in the theft and denied knowing about the other person's intent to steal. Gabriela is a Romanian national and does not have a permanent residence in Finland. The trial would thus require recourse to international legal aid and would probably require considerable resources. If found guilty, the expected penalty would be a mild fine. The store considers that the cost of

pursuing the case is clearly disproportionate to the penalty that might be expected and to the severity of the case, which is not of important public interest. The store will therefore drop the case.

Dream big

IOANA: As I'm walking down the street with music in my headphones and confidence rushing through my body on my way to meet Gabriela, on a warm sunny spring day, I notice Gabriela sitting by the road, not exactly at the meeting spot we had agreed on. Gabriela is observing me walking in her direction, with what I perceive as suspicion or mistrust. As I see myself through her scrutinising eyes, I feel shaken by this image. I take off my headphones and sit next to Gabriela, asking her how she is, with some reticence.

GABRIELA: I'm tired and stressed. Some of the other women at work keep harassing me, telling me my work is no good, that I'll never clean as good as them, that I should stay in my place and not dream big. They think you're giving them less hours because of hiring me.

I: But I've assured them so many times that I always divide the workhours fairly.

G: How are they supposed to believe you're fair when they notice you spend time with me outside work? They think I also decide on how the work-hours are divided. They say that's unfair because they've worked here longer than me. Organise a meeting with everyone. They'll calm down after they've been heard.

I: You're right. I'll send a message to the group chat straight away.

G: In the next board meeting, tell your boss to make a new rule. That when we clean, we should stop working in pairs. We should take one floor each and work independently. That way they will not have time to harass me. Tell your boss this will improve the quality of cleaning and reduce the number of complaints from clients.

I: In the board meeting, I say what Gabriela asked of me. The board members love the proposal and agree to implement it. I feel good that Gabriela used my voice to initiate board decisions at her workplace in a racial capitalist system that denies her board membership, although I took the credit for her idea. Gabriela's subversive strategy was thus conditioned by the imperative to survive in a world that does not allow "Roma life" to "flourish" (Costache 2021).

'Liberation'?

KAMU: Hey! I'm Kamu, your robot assistant. The processing of your application for EU citizen's right of residence in Finland has begun. Your place in queue is 654. The place can stay the same, change, or even go up during the process. Applications are not always processed only based on the place in queue. We will contact you if we need any additional information.[11]

GABRIELA: After six months of waiting with no decision from Migri, I see that Ioana's computer talks with them are not enough. Whenever I ask her about my application, she types some words on the computer and tells me there are still hundreds of people in front of me. Gadje are used to have the system work for them, but we Roma know it's not like that for us, that we have to push our way through. So, I go to Migri's office in person to talk to someone. I give them my case number and tell them, "Translator. Romanian." After some time, they get one by phone. They tell me I should submit more bank statements to prove that I'm still working and earning a good enough salary. I tell them to write this in Finnish on a piece of paper. I go to the bank, show them the paper, and they give me the statements. Shortly after submitting them, I receive a letter. I tell Ioana to read it to me. The decision is positive! I now have access to all of Finland's systems. I can see a doctor, I can apply for an apartment, I can bring my children here, I can study … There are so many options for me now!

IOANA: Not only did Gabriela go to Migri by herself, her initiative also sped up the process and contributed to the positive decision. Her EU citizen's right of residence in Finland has thus finally been registered. I'm pleasantly surprised and extremely happy for Gabriela, but also a bit hurt that I was not included in the final steps of the process. I reflect on my hurt feelings. Where do they come from? Do I feel some sort of entitlement over the process just because I initiated it? What were my motivations then? Was I doing it for Gabriela, or just to feel good about myself, like someone engaging in a charity act? Was I trying to paternalistically "empower" Gabriela in my own terms? I shake off my hurt feelings and tell myself that the purpose was this, for Gabriela to achieve independence. But was she not independent before? How is this independence measured, according to whose terms? Gabriela indeed enacted a present in which her agency is no longer constrained by gadje mediators and envisioned a future in which she "doesn't just survive but actually flourishes" (Costache 2021). Yet this future vision seems to rely on Gabriela governing and "liberating" herself according to gadje norms …

Final reflections

Through the creolizing fabric of our friendship, stories, and ways of working together, we shifted, however subtly, the geographies of what is possible for each of us, without erasing the power differences between us (Monahan 2011, 206). Creolization for us emerged from the "creative and affective crossings within which our lives met and evolved," like spending time together, cooking, sharing food and drinks, entering each other's circles of close friends, witnessing each other's daily habits and encounters, and offering each other mutual advice and emotional support through the daily struggles of life, which created a "relational and transversal character of a living together" (Gutiérrez Rodríguez 2015, 84–85). Yet creolization also emerged within the logic of socio-economic (re)production and dynamics of racialisation, such as workplace hierarchies and encounters with authorities, which entail a "juncture of subjugation by and liberation from governance technologies and practices" (Gutiérrez Rodríguez 2015, 95) and a process of producing hierarchical order when we cannot help not reproducing it due to our unequal power positions (Lugones 2003, 115).

While critiquing our current worlds, bringing them in generative conversations, and envisioning desired futures, our stories also speak of a failure to imagine future worlds that "draw on markedly Romani epistemologies" to create "new images, new symbols, new myths" (Costache 2021). As Ioanida Costache describes the artistic practice of Mihaela Drăgan—Roma actress, playwright, and cofounder of Roma feminist theatre collective Giuvlipen—what is needed is "a decolonizing move of 'world-shattering' that rejects the status quo, but also goes beyond critique in forging a new, be it imagined, world of liberation for the Roma" (Costache 2021). Drăgan envisions this new world by staging via theatre and film her vision of Roma Futurism (Drăgan 2021), a time-space where witchcraft merges with new technologies "to forge futuristic utopias that reimagine and reconfigure social hierarchies of oppression" (Costache 2021).

Yet it would be a gesture of appropriation and colonisation for Ioana to envision a new world of liberation for the Roma from her privileged gadji position. What she can, and has tried to do in this chapter, is envision a new world in which gadje researchers bring collaborations with Roma participants to the next level. Furthermore, for Gabriela to envision liberation beyond having access to the rights and opportunities usually available to gadje, she should have already lived in a world where such access was a reality. This shows the importance of reparations for descendants of Roma people who have historically endured slavery and genocide, an historical legacy that affects future generations and brings about high discrepancies in the level of resources and opportunities afforded to Roma and to gadje (Matache and Bhabha 2021).

Creolization has unpredictable consequences that can only be imagined (Glissant 1997, 34). We see our stories, and particularly their complicities, tensions, misunderstandings, and disagreements, as openings from which we contest hierarchies and inequalities, and as preliminary steps toward imagining new possible worlds of Roma-gadje "creolized conviviality" (Gutiérrez Rodríguez 2015).

Notes

1 Faculty of Education and Culture, Tampere University, Finland. ioana.tistea@tuni.fi
2 The Romanes term gadje—plural of gadji (feminine)/gadjo (masculine)—refers to outsiders to Romani communities. Gadje-ness is associated with whiteness, but it does not only refer to people perceived as white, but rather to people who benefit from institutional and structural privileges grounded in white supremacy (Matache 2017).
3 Gabriela slept in an emergency accommodation centre, usually referred to by its residents as "the social." She also cleaned the centre during the day, as part of her job.

4 https://www.oodihelsinki.fi/en/
5 State socialism in Romania lasted from 1947 to 1989.
6 Letter Gabriela received from Migri shortly after applying for residence. Gabriela signed a power of attorney document allowing Ioana to read and translate the letter. Gabriela chose to include this letter in the chapter.
7 Legal advice from a lawyer whom Ioana had found through connections from activist circles. We juxtaposed the letter from Migri with the legal advice as a form of resistance. However, resistance is only the first step, while Gabriela's strategy explained in the next endnote could be a next step.
8 Gabriela's reflections as Ioana translates to her the letters from authorities. Gabriela is not speaking directly about the letters. She is speaking nearby them. Trinh T. Minh-ha uses the strategy of speaking nearby subjects as a way not to contain and seize them with a unifying, authoritative narrative, thus opening the narrative to multiple possible meanings (Chen 1992). Gabriela uses the strategy of speaking nearby authorities as a way to mock power, to subversively and creatively defy norms that subdue her (Lugones 2003, 100).
9 Pseudonym.
10 Gabriela signed a power of attorney document allowing Ioana to read and translate the letter. Gabriela chose to include this letter in the chapter.
11 Reply from Migri's chatbot when checking Gabriela's application status: https://migri.fi/en/chat1

References

Back, Les, Shamser Sinha, and Charlynne Bryan. 2012. New Hierarchies of Belonging. *European Journal of Cultural Studies* 15 (2), 139–154. https://doi.org/10.1177/1367549411432030

Bogdan, Maria, Jekatyerina Dunajeva, Tímea Junghaus, Angéla Kóczé, Iulius Rostas, Márton Rövid, and Marek Szilvasi. 2018. Introducing the New Journal Critical Romani Studies. *Critical Romani Studies* 1 (1), 2–7. https://dx.doi.org/10.29098/crs.v1i1.19

Brooks, Ethel, Colin Clark, and Iulius Rostas. 2021. Engaging with Decolonisation, Tackling Antigypsyism: Lessons from Teaching Romani Studies at the Central European University in Hungary. *Social Policy and Society* 20 (4), 68–79. https://dx.doi.org/10.1017/S1474746421000452

Chen, Nancy N. 1992. Speaking Nearby: A Conversation with Trinh T. Minh-ha. *Visual Anthropology Review* 8 (1), 82–91.

Costache, Ioanida. 2018. Reclaiming Romani-ness. *Critical Romani Studies* 1 (1), 30–43. https://dx.doi.org/10.29098/crs.v1i1.11

Costache, Ioanida. 2021. Roma Futurism and Roma Healing: Historical Trauma, Possible Futures, and a New Humanism. *REVISTA ARTA*, March 23. https://revistaarta.ro/en/roma-futurism-and-roma-healing-historical-trauma-possible-futures-and-a-new-humanism/

Drăgan, Mihaela. 2021. Roma Futurism Manifesto: Techno-witchcraft is the Future. *REVISTA ARTA*, March 23. https://revistaarta.ro/wp-content/uploads/2021/03/roma-futurism-manifesto.pdf

Du Bois, W.E.B. 2005/1906. *The Soul of Black Folk*. New York: Pocket Books.

Enache, Anca. 2020. Eastern European Roma as a New Challenge for Research. In *Mobilising for Mobile Roma: Solidarity Activism in Helsinki in the 2000s-2010s*, eds. Aino Saarinen, Airi Markkanen, and Anca Enache, 55–81. Helsinki: Osuuskunta Trialogi - TRIA and Trialogue Books.

Gay y Blasco, Paloma, and Liria Hernández. 2020. *Writing Friendship: A Reciprocal Ethnography*. Palgrave Macmillan Cham. https://doi.org/10.1007/978-3-030-26542-7

Gheorghe, Carmen, and Claudia Mocanu. 2021. *Challenging intersectionality: Roma women's voices and experiences*. http://e-romnja.ro/wp-content/uploads/2021/04/Research-Intersect-Voices-.pdf

Glissant, Édouard. 1997. *Poetics of Relation*. Trans. and ed. Betsey Wing. Ann Arbor: University of Michigan Press.

Gordon, Jane Anna, and Drucilla Cornell, eds. 2021. *Creolizing Rosa Luxemburg*. London: Rowman and Littlefield.

Gutiérrez Rodríguez, Encarnación. 2015. Archipelago Europe: On Creolizing Conviviality. In *Creolizing Europe: Legacies and Transformations*, eds. Encarnación Gutiérrez Rodríguez and Shirley Anne Tate, 80–99. Liverpool: Liverpool University Press.

Himanen, Markus. 2019. "Living in Fear": Bulgarian and Romanian Street Workers' Experiences with Aggressive Public and Private Policing. In *Undoing Homogeneity in the Nordic Region Migration: Difference, and the Politics of Solidarity*, eds. Suvi Keskinen, Unnur Skaptadóttir, and Mari Toivanen, 162–178. New York: Routledge.

Keskinen, Suvi, Aminkeng Atabong Alemanji, Markus Himanen, Antti Kivijärvi, Uyi Osazee, Nirosha Pöyhölä, and Venla Rousku. 2018. The Stopped – Ethnic Profiling in Finland, *SSKH Notat - SSKH Reports and Discussion Papers* 1. http://www.profiling.fi/wp-content/uploads/2018/04/The-Stopped_ENGL.pdf

Le Bas, Damian. 2010. The Possible Implications of Diasporic Consciousness for Romani Identity. In *All Change! Romani Studies through Romani Eyes*, eds. Damian Le Bas and Thomas Acton, 61–69. Hatfield: University of Hertfordshire Press.

Lee, Ken. 2000. Orientalism and Gypsylorism. *Social Analysis: The International Journal of Social and Cultural Practice* 44 (2), 129–156.

Lugones, María. 2003. *Pilgrimages/Peregrinajes: Theorizing Coalition Against Multiple Oppressions*. Washington, DC: Rowman and Littlefield Publishers.

Martín Sevillano, Ana Belén, and Oksana Marafioti. 2021. Foreword. *Critical Romani Studies* 3 (2), 4–8. https://doi.org/10.29098/crs.v3i2.125

Matache, Margareta. 2016. *The Legacy of Gypsy Studies in Modern Romani Scholarship*. FXB Centre for Health and Human Rights, Harvard University. https://fxb.harvard.edu/the-legacy-of-gypsy-studies-in-modern-romani-scholarship

Matache, Margareta. 2017. *Dear Gadje (non-Romani) Scholars*... FXB Centre for Health and Human Rights, Harvard University. https://fxb.harvard.edu/2017/06/19/dear-gadje-non-romani-scholars/

Matache, Margareta, and Jacqueline Bhabha. 2021. The Roma Case for Reparations. In *Time for Reparation? Addressing State Responsibility for Collective Injustice*, eds. Jacqueline Bhabha, Margareta Matache, and Caroline Elkins, 253–271. University of Pennsylvania Press. https://doi.org/10.2307/j.ctv1f45q96.19

Monahan, Michael J. 2011. *The Creolizing Subject: Race, Reason, and the Politics of Purity*. Bronx, NY: Fordham University Press.

Ryder, Andrew, Angela Kóczé, Iulius Rostas, Jekatyerina Dunajeva, Maria Bogdan, Marius Taba, Marton Rövid, and Timea Junghaus, eds. 2015. Nothing About Us Without Us? Roma Participation in Policy Making and Knowledge Production. *Roma Rights Journal* 2. http://www.errc.org/uploads/upload_en/file/roma-rights-2-2015-nothing-about-us-without-us.pdf

Selling, Jan. 2018. Assessing the Historical Irresponsibility of the Gypsy Lore Society in Light of Romani Subaltern Challenges. *Critical Romani Studies* 1 (1), 44–61. https://dx.doi.org/10.29098/crs.v1i1.15

Sinha, Shamser, and Les Back. 2014. Making Methods Sociable: Dialogue, Ethics and Authorship in Qualitative Research. *Qualitative Research* 14 (4), 473–487. https://doi.org/10.1177/1468794113490717

Soares, Luiz Eduardo, MV Bill, and Celso Athayde. 2005. *Cabeça de Porco*. Rio de Janeiro: Editora Objetiva.

Tervonen, Miika, and Anca Enache. 2017. Coping with Everyday Bordering: Roma Migrants and Gatekeepers in Helsinki. *Ethnic and Racial Studies* 40 (7), 1114–1131. https://doi.org/10.1080/01419870.2017.1267378

Todorova, Miglena. 2018. Race and Women of Color in Socialist/Postsocialist Transnational Feminisms in Central and Southeastern Europe. *Meridians* 16 (1), 114–141. https://doi.org/10.2979/meridians.16.1.11

Turda, Marius. 2010. *Modernism and Eugenics*. London: Palgrave Macmillan.

Vergès, Françoise. 2021. *A Decolonial Feminism*. London: Pluto Press.

8 Decoloniality

Between a travelling concept and a relational onto-epistemic political stance

Madina Tlostanova

Prologue

When I first discovered for myself the concept of coloniality and the decolonial option in 1999, it was still relatively unknown outside Latin America and rather restricted US circles, and easily confused with postcolonial theory and anticolonial movement as more established phenomena. In the next 30 years, decoloniality has become increasingly well known and popular in the world, has travelled to different continents, and in the last five years has turned into a new buzzword that is being attached to any hip and at times pretentious intellectual endeavours. Everyone is decolonising everything these days. Decoloniality, decolonial thinking, decolonial option[1]—there are zillions of texts written on the topic. In parallel with this academic movement there are also numerous social movements and activist groups which use the slogan of decoloniality but understand it differently. Yet decoloniality is not a new universalist metatheory that one can attach to anything just as it is not a situational tactical slogan used by specific disenfranchised groups in their fights with the state or the corporations. Moreover, decolonial thought was shaped as a contextually specific discourse. The conditions that made possible the formulation of decolonial thinking in the first place are worth revisiting before we try to understand what potentials does decoloniality have in the future, and is it or should it be applicable in other places like Nordic Europe?

How decoloniality emerged: a historical, geopolitical, and theoretical context

A central concept of decolonial thought is "coloniality"—a special kind of imperial/colonial relations that emerged in the Atlantic world in the 16th century, and brought imperialism and capitalism together thus launching modernity as an overarching global project, with the help of racial taxonomising, management of knowledge production and distribution, shaping of subjectivities, and sexual and gender identities (Tlostanova and Mignolo 2012). The concept of coloniality is what makes the decolonial option ultimately different from other discourses dealing with colonialism, imperialism, and respective resistances. The idea was first coined by Peruvian sociologist

DOI: 10.4324/9781003293323-9

Anibal Quijano (1992) at an uneasy moment of the collapse of the state socialist system and discrediting of the socialist utopia, the last grand social utopia of the 20th century, and the arrival and assertion of neoliberal global capitalism as the only legitimate narrative on the planet.

In fact, decolonial thinking the way we know it could not possibly emerge in the previous anticolonial and Cold War era, just as it could not emerge before the postmodernist theoretical injection. The collective neoliberal West was enthusiastically celebrating its victory in the Cold War and producing shallow, short-lived but provocative slogans such as the infamous "the end of history" (Fukuyama 1992) and fantasising about the imminent global world of erased borders and happy consumers. The global left was confused and discouraged and has not been able to offer any convincing social and political models since then and has mainly preoccupied itself with lamenting the losses and rapidly becoming the phenomenon of the past 20th century. A rather lose group of mostly Latin American and diasporic researchers who came to be known as modernity/coloniality collective formulated their first ideas in this rather grim context of partial defeat and a realisation of the impossibility to immediately (or any time soon) implement ideals of equality and social justice.

Conceptually they were entering the scene when post-constructivist ways of thinking had already become normative. Decolonial voices largely fit into this general epistemic modality of discarding any universalist ideals, grand utopias, and master narratives. This made them different from the more conventional Marxist critique of capitalism (even if some of the original Latin American decolonialists considered themselves to be Marxist) as from the start decolonial thinkers were not interested in bringing to the world a new Truth with a capital letter, not imposing a new common and shared happy future for all people regardless of their differences. In fact, I believe that decoloniality only starts properly as an original and powerful critical discourse when it reworks and overcomes its own previous Marxist and theological delusions and this, for obvious reasons, happens only after 1989 when a distinctly different political and epistemic modality begins to arrive. Bitterly aware of the impossibility of reaching a decolonised condition any time soon, the initial Latin American and later additional voices coming from other regions of the world, have helped decolonial thinkers to increasingly reflect on some loosely bound, pluriversal, relational, differential, locally grounded patterns through which decoloniality has emerged as an open process and not as an attainable or even definable result (Annus 2019; Boatcă 2016; Kalnačs 2016; Karkov 2015; Stamenkovic 2015).

In this sense a decolonial refusal to concretise and spell out the elements of any future utopian society that decolonialists might be after, is not a fault but an intentional stance which nevertheless makes it vulnerable for critique. The pessimistic and negative context from which decoloniality sprang proved to be positive in the long run as it made decolonial activists and thinkers to not only regroup in the new situation but also look more deeply into the broader and more fundamental reasons behind this temporary defeat. This allowed touching upon the areas that have not been central in anticolonial discussions before, such as the production of knowledge and "aesthesis." The idea of "coloniality" reflected the disillusionment and transference of decolonisation rhetoric that was typical for the Cold War, from material political struggles to more imaginary and soft spheres such as knowledge production and aesthesis (Mignolo 2010; Tlostanova 2011). In Walter Mignolo's words it was "a response from the underside to the enforced homogeneity of neoliberal modernity and to the realization that the state cannot be democratized or decolonized" (Walsh and Mignolo 2018, 106). The latter part of this comment throws light at the initial deep decolonial disenchantment with the state as an institution, in both its post/neo-colonial and, indirectly, in its socialist versions. It also reflects an effort to transfer the struggle into the knowledge production and distribution area and later to aesthesis. These two related areas of decolonial application are the areas where it is still possible to resist even in the harsh political context of neoliberal totality. Importantly they do not directly address economics, the tactics of seizing and transferring political power. They do not prepare for a close revolution or a fight for independence with some clear positive result in mind, but rather connect and work with people`s consciousness, aiming for a slow changing of the way they think and see the world and themselves. The latter is a much longer and subtler process, but its results are evident today, 30 years after the emergence of decolonial thinking. At the same time, in these 30 years certain kinds of academic decolonialism have drifted further and further away from the actual ongoing local struggles for land, languages, and the autonomy. As a result, globally there is a growing gap between decolonial thinking and praxis[2] which is contradicting the basic decolonial premise and opens the door for the appropriation, trivialisation, and depoliticisation of decolonial discourses by various status quo and mainstream intellectual groups and research schools. In what follows I will reflect on this decolonial epistemic turn and its advantages and pitfalls.

I also argue that the emerging prototypes of the possible future decolonial communities of change that would overcome the modern/colonial division into theory and practice, would have to arrive—and are in some cases already arriving—at the cusp of the intersecting yet diverging academic and activist decolonialisms. A new generation of decolonial thinkers is increasingly active today and successfully combines the bottom-up social movement element with an excellent command on conceptual academic decolonial works, often via art activism. Decolonising the Euromodern division into theory and practice it is necessary to work on bridging the gaps between the extremes of decolonial academism and decolonial activism. Thinking should not be different from acting and doing, they are equally important for advancing the decolonial agenda and need each other equally. The local decolonial initiatives neglecting academic decolonialism risk remaining narrow and short-lived stand-point positions with no links to other similar movements, whereas academic decolonialists need to learn from and with these movements, think together with them, and also make sure that decolonial writings do not become instances of intellectual extractivism. Although this has been one of the key points in academic decolonialism, it is never enough to remind ourselves of its importance.

A decolonial shift to knowledge production and aesthesis

Elsewhere I have argued that decoloniality is attractive for so many different thinkers and activists because it critically analyses and questions the very modern/colonial mechanisms of knowledge production and distribution rather than just describing or condemning different historical versions of colonialism, racism, or classism or attempting to formulate some universal theory for these multiple and diverse cases (Tlostanova 2015a, 2015b, 2019). As one of the main moving forces of modernity, ontological othering evidently has epistemic roots because modernity arrives first as a self-legitimating knowledge generating system and not as an objective historical process. This ideational interpretation of modernity is what makes decolonial thinking different from most realist accounts, as it shifts the focus to the way historical processes are described and interpreted rather than to what they actually are or might be. Decolonial thought also claims that there is no objective knowledge. Knowledge(s) is/are always constructed by someone and in someone's interests, from a particular spatial, historical, and corporeal positionality. In its most extreme versions, academic decolonial thinking regards modernity as primarily a set of epistemic assumptions, premises, cognitive operations, disciplinary divisions, that were later ontologised and globally naturalised via familiar paths of capitalism, Christianity, racism, liberalism, neoliberalism, etc. And their denaturalisation and defamiliarisation is then regarded as one of the main decolonial tasks. Clearly, decoloniality in this version derives from the same origin as the poststructuralist critique and faces the same challenges of big promises of specific decolonial critical tools and lacking the actual original theoretical instruments for their implementation, and therefore sliding into the familiar Euromodern conceptual toolbox. Always accentuating the process and not the result in this essentialised academic decoloniality is another sign of its common conceptual roots with poststructuralism which, as Lewis Gordon suggests, makes it prone to fetishising and turning into idolatry with a typical moralistic investment (Gordon 2021, 16).

More specifically decoloniality attempts to put the usual subject/object hierarchy on its head and question the Western imperial epistemic duress which is complicit in maintaining the established knowledge production institutions and measuring rods, from a position of those who have been denied subjectivity and rationality and regarded as mere tokens of our culture, religion, sexuality, race, or gender. In this case stressing the subjective specificity of our knowledge (or in decolonial terms, the geopolitics and corpo-politics of knowledge) is different from a mere postmodernist claim at situated knowledges as it refers to our actual becoming epistemic subjects and looking at/reflecting on the world from the position of our own origins, lived experiences, and education. Decolonial critique is pluriversal rather than universal, and constantly aware of its own positionality while addressing the "hubris of the zero point," which, according to S. Castro-Gómez, is a specific Eurocentric positionality of the sensing and thinking subject, occupying a delocalised and disembodied vantage point which eliminates any other possible ways to produce, transmit, and represent knowledge, allowing for a world view to be built on a rigid essentialist progressivist model (Castro-Gómez 2007, 433).

The hubris of the zero point is the core of the Eurocentric epistemic contract that was launched as a result of a self-referential and self-legitimating epistemic system disavowing all other systems as non-belonging to modernity and therefore irrelevant (Tlostanova and Mignolo 2012). Decoding the specific manifestations of this typically modern/colonial aberration which is coding everything non-rational and non-European as nonmodern and non-pertaining to the sphere of knowledge, and suggesting the means for the elimination of this aberration, is perhaps the most important part of the decolonial agenda.

Decolonial gnosis or border thinking is a specific cognitive instrument that helps in realising the locality of Western epistemology and lets our assumptions shaped on its basis move beyond the normative models of knowledge production and dissemination. It is an epistemic response to coloniality formulated from the colonial difference and therefore escaping the totality of modernity's control. The initial impulse of decolonial gnosis is often a discrepancy between having to live in the colonial matrix and never really belonging to its memories, feelings, and ways of sensing and cognition. The gap between the corpo-politics of knowledge and perception, and the established mainstream ways of knowing, is what prompts the negotiating border thinking and acting as an in-between positionality of neither/nor or both/and.

The same delinking logic is detectable in decolonial aesthesis. Positioned at the intersection of being and knowing through the body, as an imperfect instrument of perception that mediates our cognition, decolonial aesthesis is an ability to perceive through the senses and the process of sensual perception itself—visual, tactile, olfactory, gustatory, etc. Setting aesthesis free lets us delink from the dominant politics of knowledge, being, and perception, which is grounded in the suppression of geo-historical dimensions of affects and corporalities (Tlostanova 2011). Significantly, decolonial aesthesis is not confined to the art sphere, but rather spreads over the way we sense and perceive the world, playing a key role in the knowledge production.

A decade ago, Walter Mignolo rephrased the famous Cartesian dictum "Cogito ergo sum" into "I am where I think":

> 'I am where I think' sets the stage for epistemic affirmations that have been disavowed. At the same time it creates a shift in the geography of reasoning. For if the affirmation 'I am where I think' is pronounced from the perspective of the epistemologically disavowed, it implies 'and you

> too,' addressed to the epistemology of the zero point. In other words, 'we are all where we think,' but only the European system of knowledge was built on the belief that the basic premise is 'I think, therefore I am,' which was a translation into secular terms of the theological foundation of knowledge (in which we already encounter the privilege of the soul over the body) to secular terms.
>
> (Mignolo 2011, 169)

Although radical delinking combined with the corpo- and geopolitical dimensions of knowledge production are still perfectly sound decolonial claims, what has increasingly disturbed me is how can we, decolonial thinkers, implement them. I believe that denouncing Eurocentrism decolonial thought at times tends to still paradoxically reproduce its main conceptual premises, not so much content-wise but discursively. It takes place in the still binary mode of thinking that tends to discard the important nuances for the sake of singling out the general structures and tendencies. This urge leads to homogenising and effectively demonising the collective West, often as a convenient rhetorical gesture rather than a seriously grounded argument. This tendency is also expressed in temporal models that problematise linearity yet still reproduce the progressivist historical scheme (albeit with a focus on the eulogised past). In recent decolonial texts, the previously much more anthropocentric decolonial notions are increasingly balanced with thoroughly conceptualised inclusion of nature and even life as such into the sphere of exteriority and shifting the human being to its deserved humble status (Vázquéz 2017; Walsh and Mignolo 2018). Yet the construction of arguments itself remains surprisingly Euromodern. Partly this is due to the academic writing requirements which remain logocentric if not positivist and therefore ensnare the academic decolonial thinker into a vicious loop of having to use the master`s tools to dismantle his house. This is also why non-academic decolonial writing and art activism, with their polysemantic metaphors and complex and non-straightforward semiotic nodes, are more successful in capturing decolonial sensibility and agendas than any academic text could ever hope to do.

Hijacking the decolonial jargon for colonial(ist) purposes

A focus on knowledge production and aesthesis makes decolonial critique broader and more profound than the anticolonialism of the Cold War era as it aims to question the terms and not the content of the conversation. Yet this ideational bent is also what makes decoloniality vulnerable—it is too often and easily translated into the language of mainstream critical theory, in fact, reduced to it. A mistake or intentional malice of the mainstream appropriators of decolonial thinking is largely that they hijack its frame and main concepts and terms, yet take themselves out of the thinking process, effectively delocalise their argument in the familiar Eurocentric mode, and never stress in what capacity are they applying decolonial thinking. Therefore, one of the disturbing developments in decoloniality is its hollowing out, trivialisation, and appropriation by generally mainstream status quo thinkers who tend to confuse it with a delocalised critique of rationalism, logocentrism, progressivism, and modernity.

A different but no less misleading confusion arrives when decoloniality is eroded and merged with some generalised anti-racist and anticolonial rhetoric that might have a completely different genealogy, ideology, and philosophy (Liinason and Alm 2018; Liinason and Cuesta 2016). Two examples of this tendency can be found in recent texts by Leon Moosavi (2020) and Sabelo J. Ndlovu-Gatsheni (2021) which I regard as indications of the ongoing blurring of the decolonial paradigm and its turning into an all-encompassing boundless anything-goes discourse where the word "decolonial" is increasingly metaphorical rather than strictly terminological. Moosavi`s article is a review work, or perhaps a selective analytical compilation that joins the rising critique of decoloniality as a new buzzword, yet fails to differentiate between the anticolonial, postcolonial, and decolonial discourses and their respective historical and geopolitical genealogies, dynamic contexts, and mutual relations, which is evident in the way the author confuses or neglects to separate decolonisation and decoloniality. Paradoxically, the author himself attempts to jump into the decolonial bandwagon, although from the position of the South. It is less ethically and politically problematic than the Northern appropriation, but it still leads to a blurriness and homogenising of categories and concepts.

Similar tendencies are evident in a much more decolonially informed work by Ndlovu-Gatsheni that offers an excellent genealogy of Africana anticolonial critical thought and agency yet for some reason still attempts to write it into the decolonial discourse and claim its space and visibility there. Once again, a very specific terminological meaning of decoloniality is replaced with a blurred and boundless metaphoric interpretation. I am also uncomfortable with the implied principle of inclusion—this time not to the mainstream Euromodern canon but to a fashionable decolonial one. As in Moosavi`s case, here too decolonisation and decoloniality are confused indicating a general move to indiscriminate lumping together the anticolonial, postcolonial and decolonial positionalities.

More and more initiatives, projects, conferences apply the term "decolonial" or "decolonisation." There is an increasingly large number of efforts to decolonise museums, universities, public institutions, as well as many other aspects of life such as sexualities (Bakshi et al. 2016), diets (Mailer and Hale 2017), nature (Demos 2016), design and urban spaces (Kalantidou and Fry 2014). However, in some cases the use of decoloniality is not entirely justified. The problem is not only the lack of actual knowledge of the genealogies of decolonial ideas and struggles, but also a depoliticisation of the originally radical decolonial paradigm and its lumping together with, first of all, postcolonial theory. On a closer inspection, many recent decolonial events appear no more than a neoliberal epistemic appropriation. Decolonisation becomes a buzzword abused in the titles of books, conferences, and research initiatives which in many cases are detached from the reality of decolonial struggles in the peripheries of Europe, among the indigenous people or in the Global South. It is especially true of European countries, where this discourse has arrived with considerable delay except for the marginalised events organised by a handful of decolonial thinkers themselves or their few European counterparts.[3] But all these seminars, summer schools, conferences either remained for a long time in the periphery of cutting-edge European-style modern/colonial research and therefore remained ignored and invisible on a larger mainstream scale, or were quickly remade and restructured to fit the more conventional postcolonial, multiculturalist, and generally highly commercial modes. It is only in the last few years that the word "decolonial" has started to dominate the mainstream conference and research European environments in highly problematic ways (8th Conference on the New Materialisms 'Environmental Humanities and New Materialisms: The Ethics of Decolonizing Nature and Culture', Paris 2017; Decolonizing transgender in the North—4th Nordic transgender studies symposium, Karlstad 2016; Decolonizing North Conference and Exhibition, Stockholm 2017). Decoloniality has been also appropriated by mainstream art institutions and theories, losing its element of contestation, and turning into nicely packaged and easily digestible postcolonial goods treated through familiar Orientalism, exoticisation, demonisation, and other such Eurocentric knowledge frames (McClintock 1992). Even more disturbing is the fact that the ultra-right rhetoric has hijacked the decolonial readings of the past as its core agenda, to apply it in a dangerous reanimation of ethnic-cultural, religious, Eurocentric, nationalist, and other mythologies normalising rootedness and othering non-belonging (Morozov 2015; The Return of the Colonial 2020; Tlostanova 2018).

In some sense this European decolonial trajectory is reminiscent of an earlier example of intersectionality which has gone from a radical black feminist stand-point discourse (Hull et al. 1982) to a blurred and depoliticised reinterpretation within contemporary European mainstream feminism where it has become a position of belonging to some vague common global transnational feminist culture (Carbin and Edenheim 2013). As in the case of intersectionality then it is crucial to ask—who speaks in and of decoloniality in Europe today and from what position is the enunciation made? Who enunciates decoloniality? Is this enunciation a birth of a new discipline—decolonial studies? And in what intersection of decoloniality does the enunciation take place? Why is it often the case that the European discussants of decoloniality stand above the issues they discuss as the observers and remain untouched by the intersections and power asymmetries in question? It is much more important to focus on different tangential genealogies of knowledge, being, gender, perception, and, once again, to shift the emphasis from the enunciated to the enunciation.

Whitewashed and sanitised "decolonial studies" that fail to see the profound differences between postcolonial theory and decoloniality and often substitute decoloniality for deconstruction yet keep the Euromodern epistemic framework intact, is what we find 30 years after the launch of the decolonial option in European—and especially in the Nordic—contexts. The latter requires specific attention as they somehow differ from other European responses to decoloniality. Therefore, in the next section I will briefly address some of the specifically Nordic issues with decoloniality.

Decoloniality in the Nordic context

One of the main problems in the Nordic context is the way postcolonial and decolonial paradigms are consistently mixed and confused by many researchers without realising their fundamental epistemic differences. One example is the Linnaeus University Centre for Concurrences in Colonial and Postcolonial Studies, which announces on its webpage that it aims at a more "balanced and empirically sound historiography of global encounters throughout modern history" in which the non-European others will be presented as active participants of "concurrent" i.e. ultimately competitive relations (Centre for Concurrences 2021). The concept of concurrencies is disturbing from a decolonial perspective. Claiming anticolonial justice the concurrencies model unproblematically reproduces the main aberration of modernity—that of agonistics, additionally erasing or muddling the power asymmetries involved. Through an essentially colonial prop of comparative studies, it hides the god-trick positionality, and turns the drama of enforced and homogenising coloniality into a false narrative of many coexisting and competing powers. What is more is that this text lumps together the hastily invented colonial studies with the appropriated postcolonial ones, thus erasing even a small critical element in the postcolonial paradigm that has been linked with the geopolitics and corpo-politics of the postcolonial researchers themselves. This ultimately limits the postcolonial once again via its narrow temporal understanding. In other words, this approach reinstates epistemic coloniality with its characteristic subject/object division and fails to problematise the involved (predominantly white and privileged) researchers as producers of knowledge and their position within modernity/coloniality. Such an approach lacks a pluritopical (Mignolo 1995) or multispatial (Tlostanova 2017) hermeneutics that could help us understand something which does not belong to our horizon through a dialogic and experiential (not merely interpretative) learning from the other. In such a multispatial model, the understanding subject is placed in a colonial periphery or imperial semi-periphery, in a non-European tradition, or other marginalised space, disturbing the habitual Western vantage point and questioning the position and homogeneity of the understanding subject. Refusing to make a preference for either cultural relativism or multiculturalism, multispatial hermeneutics accentuates the politics of the embodiment and the construction of space for formulating and expressing one's active positionality. Although the understanding subject should presume the truth of what he, she, or they claim to understand, this subject should also admit the existence of alternative politics of space with equal claims to truth. Multispatial hermeneutics is grounded in relativism in the understanding of cultural and epistemic differences, yet this relativism is always written into the complex matrix of intersectional power asymmetries which means that it is always aware of the geopolitics and corpo-politics of being, thinking, and sensing, that it takes into account the so-called positionality as a specific genealogy of subject formation and its political and ethical stance. The latter, however, may evolve into a problem and also shows genealogical links with poststructuralism as convincingly argued by Lewis Gordon:

> Poststructuralism functions within decoloniality as a colonial element or form of coloniality … For decoloniality this problem becomes acute where theory is undertheorized. Where this is so the result is often an appeal to theorists with the addition of a *position* or an issue. That position, often formulated as positionality—is often a moral one offered as a political intervention.
>
> (Gordon 2021, 15)

But even if Gordon is right in pointing to the hidden pitfalls of decolonial sliding into moralism instead of politics, multispatial hermeneutics is still relevant as it rebels against the totalitarianism of the monotopical model and intends to let the others speak, reason, argue, and create as equals to the same, and from their own body and experience, thus subsuming the imperial reason that taxonomised them as others. This approach is alien to most postcolonial studies researchers and centres in the Nordic countries. Under the token inclusion of certain topics, they are largely marked by a blindness towards their own specific colonial trajectories (and especially the imperial difference) and the complexity of the struggles as well as the diversity of indigenous peoples that I will briefly address below.

As we have argued with my co-authors Suruchi Thapar-Björkert and Ina Knobblock (2019), in an effort to answer the question of Sweden needing decolonial feminism, one of the recurrent themes discussed by the Nordic researchers in their struggles to define the role and place of the Nordic countries in the larger global imperial-colonial project or "the production of Europe" according to Loftsdottir and Jensen (2016, 1), is the so-called "colonial complicity." This is a term coined earlier by Gayatri Chakravorty Spivak and appropriated by the Nordic researchers who identify themselves as postcolonial (Ipsen and Fur 2009). Keskinen, Tuori, Irni, and Mulinari (2009) argue that the Nordic countries have been complicit in (post)colonial processes through a specific construction of national imaginaries and racialisation that have been closely tied to the Nordic welfare state models and notions of gender equality. These authors tend to see the persistent structural inequalities and racialised exclusions as a residue from the colonial period that is revamped and reimposed onto different marginalised groups in the Nordic countries, most recently onto migrants and refugees. Their interventions arguably have started to destabilise the previous homogenous image of Nordic—and particularly Swedish—colonial exceptionalism (Molina 2004; Mulinari and Neergard 2017). Yet, as we argued, the postcolonial rhetoric is interpreted by the Nordic scholars in specific ways, often far from the initial interventions coming from the postcolonial subjects in the Global South, while the positionality of the Nordic postcolonialists themselves within the coloniality of knowledge is rarely taken into account or critically accessed. What is even more disturbing is the persistent conflation of postcolonial studies and decolonial thinking and/as agency which goes hand in hand with the chronic invisibilising of the discussions and contributions brought into the Nordic countries by decolonial scholars from elsewhere.

One recurrent issue with many Nordic interpretations of the imperial and colonial problematic is a lack of clear delineation between colonialism and coloniality—the core issue in trying to differentiate postcolonial studies from decolonial option. Even in their most critical versions, postcolonial studies remain within the established disciplinary mode in which a study presupposes a firm subject/object division. A successful and quick institutionalisation of postcolonial studies has required a sacrifice of choosing the side of the studying subject, not the studied object. The institutional disciplinary frame coded by the word "studies" does not presuppose, by definition, putting theory and life-world on the same axis and practising decolonisation in our everyday writing, thinking, and activism. This does not mean that postcolonial theorists neglect the corpo-politics and the geopolitics of knowledge, being, and perception, or that they do not take radically decolonising positions as activists-cum-theorists, it just means that their discipline does not require this kind of move on their part and it becomes a matter of a personal decolonial choice. As a result, what we mostly find in the case of Nordic scholars who call themselves postcolonial and recently have started to pick up some decolonial terms (Martinsson and Mulinari 2018), is that they strive to analyse coloniality through a postcolonial lens. They often fail to differentiate between coloniality as the darker side of modernity and an ongoing trace of epistemic biases and descriptive attributes of particular colonialisms. Applying decoloniality not as a mere replacement of colonialism but a totally different and independent concept would require a much more radical delinking and self-critique on their part, as well as venturing into the epistemic areas that postcolonial studies have never questioned before. As a result, the researchers are taken outside of their own argument, and remain unaware of their own (often complicit) positionality, thus reinstating the Euromodern pattern of representation and inclusion of the other and its interpretation in the language of the same, rather than a decolonial delinking from this logic.

Therefore, the belated Nordic toying with the decolonial is so far unconvincing as it misses the main point of decolonial critique which rejects representational models of telling *about* the other rather than thinking and acting *together* with other others. Hence comes the Nordic confusion of the postcolonial and the decolonial and a lack of focus on the crucial issues of situatedness and problematisation of representationalism. Additionally, not many Nordic postcolonial works testify to an awareness of the specific contextuality and historical locality of the postcolonial theory and its main concepts. In other words, it is not enough—and is at times misleading—to just borrow certain postcolonial or decolonial terms and apply them to the Nordic local histories. The Anglophone and Francophone postcolonial mainstream is characterised by its specific, locally bound (though disguised as universal) frame, context, and conditions. It is marked by with such specific features as the overseas colonies, the clearly racially marked colonial and postcolonial others and today their contemporary descendants such as Afropeans, Middle Eastern, or Somalian refugees, but almost never (until recently) the local indigenous people. Importantly, the specific colonial trajectories and grounds of racial formations and divisions in the Nordic region itself, and the internal regional factors crucial for the shaping of human taxonomies and ideas of racial and national superiority and exceptionality, have hardly been part of the Nordic postcolonial discussions until recently, as if Sweden or other Nordic countries have not existed before the emergence of their celebrated national welfare state models and all the conflicting previous histories were erased and forgotten.

Primarily this refers to the struggles of indigenous people and, first of all, the Sami, and the erased histories of the Nordic settler and internal colonialism. Thus, the numerous historical studies of Swedish and Danish colonial expansion, the *Dominium Maris Balticith* and the "Stormaktstiden" (Naum, Nordin 2013), seem to exist exclusively within descriptive factual historiography, and separately from any decolonial interventions and conceptualisations that operate with completely different notions and assumptions and therefore come to different conclusions. As Pernille Ipsen and Gunlög Fur accurately pointed out in their introduction to the special issue of *Itinerario* devoted to Scandinavian colonialism (2009),

> in general, historical writing has paid little attention to modern postcolonial dilemmas or theory. In fact, we suggest that popular historical discourses in Scandinavia have moved directly from no colonialism to post-colonialism without stopping at a thorough investigation of Scandinavian participation in and gains from colonial expansion and exchanges from the early modern period until the present.
>
> (Ipsen and Fur 2009, 10)

To bridge this still prominent gap seems to be an important task for decolonial Nordic thought which is itself yet to be born. One of the most important elements of this specifically Nordic imperial-colonial configuration and trajectory is the unique model of its imperial difference that I will briefly trace below.

The imperial difference ... Nordic style

Imperial difference and inter-imperiality (Boatcă 2012; Doyle 2013; Tlostanova 2003, 2018) as a complex and heterogeneous imperial hierarchy in modernity, have not received sufficient attention even within decolonial thinking itself, as decoloniality has mostly focused on the colonial difference as a site of knowledge production and decolonial intervention. When we shift from the first-class modern capitalist empires, such as Great Britain or France, to make sense of the second-rate imperial entities that failed to win and carve a better position for themselves in the world system yet had to compete and struggle to keep their position in the global imperial hierarchy, we have a variety of resulting models. One is exemplified by Spain, Portugal, and Italy who, having lost to their more powerful rivals, became the so-called South of Europe—largely an internal European inferior, if not a colonial space (Tlostanova and Mignolo 2012, 10–11). The Russian empire, Soviet Union, and the present Russian Federation represent a case of the external imperial difference rooted in a non-European religious, linguistic, economic, and ethnic-racial model, that in the post-Enlightenment modernity was destined for the outsider peripheral roles of the subaltern or a Janus-faced empire (Tlostanova 2003, 2008, 2015a).

The second-class modern empires mostly compete among themselves, rarely attempting to break through to get to the first imperial league (as the USSR tried and failed to do). This background of imperial drama, with its many characters that combined the imperial aspirations with inferiority complexes, dispersed, blurred, and indirect relations to the normative colonialist tools such as trade and missionary activism, further complicates modernity/coloniality, including its inter-imperial relations, and destabilises the habitual assumptions that are to be found in postcolonial and decolonial investigations such as a homogenised West, a no less homogenised or generalised idea of the other, conventional definitions of the imperial power, and subaltern resistance. In my view, it is through this frame of the imperial difference that the imperial-colonial history of the Nordic region could be complicated and revisited.

Sweden seems to be an interesting example of the internal imperial difference. Like the South of Europe, the North of Europe shares the main cultural, religious, ethnic, economic, and linguistic affinity with the European core, and therefore its imperial difference is always incomplete and partial. Agreeing with the shared Western beliefs, the official Swedish position, as several Nordic researchers of colonial histories accurately point out (Bryden, Forsgren, Fur 2017; Fur 2013), indeed stubbornly refuses to regard Swedish history as imperial. However, it is not enough to just mention the Swedish complicity or tacit agreement with the crimes of the first imperial league. It is also important to reflect on the larger picture of the global coloniality and the marginal—yet clearly imperial and superior—roles Scandinavian countries have played in this complex phenomenon as the above-mentioned special issue of *Itinerario* attempted to do as early as 2009.

Here are just some of the important elements in the specific Swedish trajectory of imperial difference that contemporary Sweden often chooses to forget in its rewriting of national history and self-image in a more attractive and positive way: the shrinking and early abandonment of the original overseas imperial appetites; their transformation into more hidden and complicit forms of trading in slaves and colonial goods; the shift to the ill-fated

Dominium Maris Balticith project marked, among other things, by the rarely addressed histories of the Swedish "benevolent colonialism" in Estonia (Tarkiainen and Tarkiainen 2013); its expansion in and annexation of Finland; the long-going rivalry for the Baltic territories with the future parts of the Czarist Empire and mainly the Novgorod Republic, that had not only an obvious mercantile but also a religious element (several Swedish crusades, Christiansen 1997) of converting the Orthodox Christian and pagan populations of the Baltic North-East into Catholicism; the massive dispossession of the Sami people of their land in the interests of settler nation-building and, later, industrialisation; and still rarely addressed histories of ethnic minorities and racialised European indigenous groups that happened to be caught in between the Swedish and Russian imperial rivalries and almost vanished as a result of their forced assimilating—and, at times, annihilating and relocating—tactics in both empires (Kurs 1994; Strogalschikova 2014), as well as the never-ending indigenous resistance and re-existence (Albán Achinte 2009). These groups include not only the better-known Sami and Ingermanland Finns, but also the forgotten and erased Vodians, Izhorians, Karelians, Vepsians, and many others. What is important to explore in the future is the actual ways and forms of the Swedish—and broader Nordic—compensation and re-channelling of the early suspended expansionist projects and internal European settler colonial schemes, that take place, once again, most easily in the epistemic and symbolic realms. This would not be applying decolonial thinking to the Nordic region but rather reflecting on how modernity/coloniality takes specific local shapes in the case of Scandinavian countries which, no doubt, will enrich decolonial thought and make it more nuanced and complex.

Taking decoloniality further, or let us make that other world possible

The examples addressed above are only one possible way to framing a decolonial lens in the analysis of Nordic history and contemporaneity which also potentially enriches decoloniality and prevents it from becoming a bronze monument to itself. Moreover, it is not only a historical reconsideration that is important for keeping decoloniality alive and vibrant, but also a more pronounced shift towards contemporary challenges that clearly go beyond the original decolonial focus on the intersection of race and capitalism, and incorporating more intensively the present challenges of climate change, increased chronophobia, defuturing, the changing human ontology (due to technological colonisation among other factors that leads to specific forms of rebranded racism and discrimination) and global unsettlement in which the geopolitical and colonial elements are accompanied by other overarching factors. In our urge to reclaim the rights for the past, for erased memories, in our struggle with the limitations of the presentist, lateral thinking, that "insists on immanence against history's melancholia" (Broeck 2018, 179), decolonialists seem to have overdone it and shut ourselves out of any discussions of the present and the future, in fact, leaving them to the very people that are marked by presentism and immanence (Vázquéz 2017), a lack of political imagination, and a reliance on technocratic illusions of the privileged. The fixation on the (resurrection and reclaiming of the) past often prevents decoloniality from imagining the future and from detecting the tectonic shifts the world is rapidly going through. Therefore, not forgetting about the past decolonial community should also turn more actively to the present and to the ways of imagining a decolonial future. Engaging with the present and the future decolonial thinking could also start bridging the gap between academic decolonialism and decolonial agency and bottom-up activism. Today's situation urges decoloniality to move in the direction of relational agency unlimited to colonial difference alone avoiding both the extreme of imagined indigeneity and a confinement to the ivory academic tower.

This balancing should be constantly aware of the pitfalls of excessive stand-pointism (usually that of colonial difference) which may divide humanity in a potentially essentialist way. It is particularly misleading in the present context of the global and complex crisis as it closes the possibility for communal refuturing agency that is necessary for our survival as a species and of other species on Earth. Such an agency should be grounded in the principle of deep coalitions defined by Maria Lugones as follows: "Deep coalitions never reduce multiplicity, they span across differences. Aware of particular configurations of oppression, they are not fixed on them, but strive beyond into the world, towards a shared struggle of interrelated others" (Lugones 2003, 98). Such horizontal coalitions require maintaining complexity and heterogeneity rather than taking them to homogenous sameness on both universalised global and/or particularised local grounds.

At the moment, voices challenging the existing framework are many, but they are still compartmentalised and isolated from each other and fight their losing battles one on one with the Euromodern knowledge production system. What is needed is that various others discarded from modernity start to nurture a genuine interest in each other's ways of thinking and being, a drive to engage with each other's ideas bypassing the Euromodern endorsement. The sad incapacity to connect transversally is a sign of the successful modern/imperial divide and rule politics, generating inferiority complexes that are reproduced generation after generation within the catching-up logic. It is time we disassembled this frame onto which knowledge, or what is considered as such, is being planted rather than merely adding new information to the old carcase. The question is if decolonial infiltrations from within the exhausted modernity are enough, or is something more radical and strategic is needed? What is obvious is that decolonial thinkers should become more active in launching the change communities of redirective action (Escobar 2017) and refuturing (Fry and Tlostanova 2020) that would incorporate different actors from the Global North and South, and from the semi-peripheral spaces—scholars, designers, activists, members of the local communities, and artists, who would work together to make the possible other world come true.

Notes

1 Recent appropriators of decoloniality have even started to nonsensically call it decolonial theory although decolonial thinkers and activists have long been saying that theory is yet another modern/colonial concept from which it is necessary to delink. For a summary of these discussion see Walsh and Mignolo, 2018.

2 Examples abound not only in Latin America and especially among the Latin American and Latinx scholars in the US, but also in Russia, Central Asia, and Eastern Europe, see e.g. Silvia Rivera Cusicanqui`s critique of academic decolonialism and its appropriation of her ideas (Cusicanqui 2012); the original and powerful re-existent decolonial artistic and/as activist agency in Kazakhstan as opposed to derivative, bleak and belated academic concoctions of postcolonial, orientalist and some decolonial ideas by academic circles in the post-Soviet space (Suleimenova 2019 Tlostanova 2020).

3 Here the *Decoloniality Europe* initiative is particularly worth mentioning as a positive example, as well as several conferences organised by the proponents of decolonial thinking in Europe.

References

Albán, Achinte Adolfo. 2009. "Artistas Indígenas y Afrocolombianos: Entre las Memorias y las Cosmovisiones. Estéticas de la Re-Existencia." [Indigenous and Afro-Colombian Artists: Between Memories and Cosmovisiones. Aesthetics of Re-Existence.] In *Arte y Estética en la Encrucijeda Descolonial* [Art and Aesthetics at the Decolonial Crossroads], edited by Zulma Palermo, 83–112. Buenos Aires: Del Siglo.

Annus, Epp. 2019. *Soviet Postcolonial Studies. A View from the Western Borderlands*. London: Routledge.

Bakshi, Sandeep, Suhraiya Jivraj and Silvia Posocco (eds.) 2016. *Decolonizing Sexualities: Transnational Perspectives, Critical Interventions*. Oxford: Counterpress.

Boatcă, Manuela. 2012. "Catching Up with the (New) West. The German "Excellence Initiative," Area Studies, and the Re-Production of Inequality", *Human Architecture: Journal of the Sociology of Self-knowledge* X, Issue 1, Winter: 17–30.

Boatcă, Manuela. 2016. *Global Inequalities Beyond Occidentalism*. London: Routledge.

Broeck, Sabine. 2018. *Gender and the Abjection of Blackness*. New York: Suny Press.

Bryden, Diana, Peter Forsgren and Gunlög Fur. 2017. *Concurrent Imaginaries, Postcolonial Worlds. Toward Revised Histories*. Netherlands: Brill.

Carbin, Maria and Sara Edenheim. 2013. "The Intersectional Turn in Feminist Theory: A Dream of a Common Language?" *European Journal of Women Studies* 20(3): 233–248.

Castro-Gómez, Santiago. 2007. "The Missing Chapter of Empire: Postmodern Reorganization of Coloniality and Post-Fordist Capitalism", *Cultural Studies* 21(2–3): 428–448.

Christiansen, Eric. 1997. *The Northern Crusades*. London: Penguin.

Cusicanqui, Silvia Rivera. 2012. "Ch'ixinakax utxiwa: A Reflection on the Practices and Discourses of Decolonization", *The South Atlantic Quarterly* 111(1), Winter: 95–109.

Decolonizing North. 2017. Konsthall and CEMFOR, Stockholm, 7–8 December. http://www.decolonizing.ps/site/wp-content/uploads/2017/11/DN.pdf

Demos T.J. 2016. *Decolonizing Nature. Contemporary Art and the Politics of Ecology*. Berlin: Sternberg Press.

Doyle, Laura. 2013. "Inter-Imperiality. Dialectics in a Postcolonial World History", *Interventions. International Journal of Postcolonial Studies* 16(2): 159–196.

Escobar, Arturo. 2017. *Designs for the Pluriverse*. Durham: Duke University Press.
Fry, Tony and Madina Tlostanova. 2020. *A New Political Imagination. Making the Case*, London: Routledge.
Fukuyama, Francis. 1992. *The End of History and the Last Man*. New York: Free Press.
Fur, Gunlög. 2013."Colonialism and Swedish History: Unthinkable Connections?", In: *Scandinavian Colonialism. The Rise of Modernity*, edited by Magdalena Naum and Jonas M. Nordinw, 17–36. New York: Springer.
Gordon, Lewis R. 2021. *Freedom, Justice, and Decolonization*. London: Routledge.
The Return of the Colonial: Understanding the Role of Eastern Europe in Global Colonisation Debates and Decolonial Struggles. Workshop: 10 September 2020. A summary and future roadmap. https://eprints.soas.ac.uk/34222/1/Workshop%20Summary%20and%20Roadmap_17.11.2020.pdf
Hull, Akasha Gloria T, Patricia Bell Scott, and Barbara Smith. 1982. *All the Women Are White, All the Blacks Are Men, But Some of Us Are Brave*. New York: Feminist Press.
Ipsen, Pernille and Gunlög Fur. 2009. "Introduction: Special Issue on Scandinavian Colonialism", *Itinerario: International Journal on the History of European Expansion and Global Interaction* 33: 7–16.
Kalantidou, Eleni and Tony Fry. eds. 2014. *Design in the Borderlands*. London: Routledge.
Kalnačs, Benedikts. 2016. "Comparing Colonial Differences: Baltic Literary Cultures as Agencies of Europe's Internal Others", *Journal of Baltic Studies* 47(1): 1–16.
Karkov, Nikolai. 2015. "Decolonizing Praxis in Eastern Europe: Toward a South-to-South Dialogue", *Comparative and Continental Philosophy* 7(2): 180–200.
Keskinen, Suvi, Salla Tuori, Sara Irni and Diana Mulinari. 2009. *Complying with Colonialism: Gender, Race and Ethnicity in the Nordic Region*. Aldershot: Ashgate.
Kurs, Ott. 1994. "Ingria: The Broken Land Bridge Between Estonia and Finland", *GeoJournal* 33: 107–113.
Liinason, Mia and Erika Alm. 2018. "Ungendering Europe: Critical Engagements with Key Objects in Feminism", *Gender, Place and Culture* 25(7): 955–962.
Liinason, Mia and Marta Cuesta. 2016. *Hoppets politik: Feministisk aktivism i Sverige idag* [The politics of hope: Feminist activism in present-day Sweden]. Göteborg: Makadam.
Linnaeus University Centre for Concurrences in Colonial and Postcolonial Studies. 2021. https://lnu.se/en/research/searchresearch/linnaeus-university-centre-for-concurrences-in-colonial-and-postcolonial-studies/
Loftsdottir, Kristín and Lars Jensen. Eds. 2016. *Whiteness and postcolonialism in the Nordic region: Exceptionalism, migrant others, and national identities*. London: Routledge.
Lugones, María. 2003. *Pilgrimages/Peregrinajes: Theorizing Coalition Against Multiple Oppression*. New York: Rowman and Littlefield Publishers.
Mailer, Gideon and Nicola Hale. 2017. *Decolonizing the Diet. Nutrition, Immunity, and the Warning from Early America*. London and New York: Anthem Press.
Martinsson, Lia, and Diana Mulinari. Eds. 2018. *Dreaming Global Change, Doing Local Feminisms: Visions of Feminism, Global North/Global South Encounters, Conversations and Disagreements*. London: Routledge.
McClintock, Anne. 1992. "The Angel of Progress: Pitfalls of the Term Post-Colonialism", *Social Text* 31/32: 84–98.

Mignolo, Walter. 1995. *The Darker Side of the Renaissance*. Ann Arbor: University of Michigan Press.

Mignolo, Walter. 2010. "Aiesthesis Decolonial", *Calle 14*, 4(4): 11–25.

Mignolo, Walter. 2011. "I am Where I Think: Remapping the Order of Knowing", In *The Creolization of Theory*, edited by Lionnet, Françoise and Shu-mei Shih, 159–192. Durham & London: Duke University Press.

Molina, Irene. 2004. Intersubjektivitet och intersektionalitet för en subversiv antirasistisk feminism [Intersubjectivity and intersectionality for a subversive antiracist feminism]. *Sociologisk Forskning* 3: 19–24.

Moosavi, Leon. 2020. "The Decolonial Bandwagon and the Dangers of Intellectual Decolonization", *International Review of Sociology* 30(2): 332–354.

Morozov, Vyacheslav. 2015. *Russia's Postcolonial Identity. A Subaltern Empire in a Eurocentric World*. London: Palgrave Macmillan.

Mulinari, Diana and Anders Neergard. 2017. "Theorising Racism: Exploring the Swedish Racial Regime", *Nordic Journal of Migration Research* 7(2): 88–96.

Naum, Magdalena and Jonas M. Nordin. eds. 2013. *Scandinavian Colonialism and the Rise of Modernity: Small Time Agents in a Global Arena*. New York: Springer.

Ndlovu-Gatsheni, Sabelo J. 2021. "The Cognitive Empire, Politics of Knowledge and African Intellectual Productions: Reflections on Struggles for Epistemic Freedom and Resurgence of Decolonisation in the Twenty-first Century", *Third World Quarterly* 42(5): 882–901.

Quijano, Anibal. 1992. "Colonialidad y modernidad/racionalidad", *Peru Indigena* 13(29): 11–20.

Stamenkovic, Marko. 2015. *Suicide Cultures: Theories and Practices of Radical Withdrawal*. PhD thesis. University of Gent.

Strogalschikova, Zinaida. 2014. *Vepsy. Ocherki Istorii i Kulturi*. [Vepsians. Sketches on history and culture]. Saint-Petersburg: Inkeri.

Suleimenova, Saule. 2019. "Khudozhnitsa Saule Suleimenova: o tom, pochemy vse sovremennie khudozhniki – aktivisti" (Artist Saule Suleimenova: on why all contemporary artistsareactivists"),STEPPE,February4,https://the-steppe.com/lyudi/hudozhnica-saule-suleymenova-o-tom-pochemu-vse-sovremennye-hudozhniki-aktivisty

Tarkiainen, Kari and Ülle Tarkiainen. 2013. *Provinsen Bortom Havet: Estlands Svenska Historia 1561–1710*. Stockholm: Atlantis.

Tlostanova, Madina. 2003. *A Janus-Faced Empire. Notes on the Russian Empire in Modernity Written from the Border*. Moscow: Blok.

Tlostanova, Madina. 2008. "The Janus-faced Empire Distorting Orientalist Discourses: Gender, Race, and Religion in the Russian/(post)Soviet Constructions of the Orient", *Worlds and Knowledges Otherwise. A Web Dossier* 2: 2, https://globalstudies.trinity.duke.edu/sites/globalstudies.trinity.duke.edu/files/file-attachments/v2d2_Tlostanova.pdf

Tlostanova, Madina. 2011. "La aesthesis trans-moderna en la zona fronteriza eurasiatica y el anti-sublime decolonial", *CALLE14* 5(6): 10–31.

Tlostanova, Madina. 2015a. "Can the Post-Soviet Think? On Coloniality of Knowledge, External Imperial and Double Colonial Difference", *Intersections. East European Journal of Society and Politics* 1(2): 38–58.

Tlostanova, Madina. 2015b. "Between the Russian/Soviet Dependencies, Neoliberal Delusions, Dewesternizing Options, and Decolonial Drives", *Cultural Dynamics* 27(2): 267–283.

Tlostanova, Madina. 2017. *Postcolonialism and Postsocialism in Fiction and Art: Resistance and Re-Existence*. Cham: Palgrave Macmillan.

Tlostanova, Madina. 2018. *What Does it Mean to be Post-Soviet? Decolonial Art from the Ruins of the Soviet Empire*. Durham: Duke University Press.

Tlostanova, Madina. 2019. "Border Thinking/Being/Perception. Towards a Deep Coalition Across the Atlantic", In: *Speaking Face to Face. The Visionary Philosophy of María Lugones*, edited by DiPietro, Pedro, Jennifer McWeeny and Shireen Roshanravan, 125–143. Albany: State University of New York Press.

Tlostanova, Madina. 2020. *Dekolonialnost Znania, Bytia i Oschuschenia* [Decoloniality of Knowledge, Being and Sensing], Almaty: Center for Contemporary Culture "Tselinny".

Tlostanova, Madina and Walter Mignolo. 2012. *Learning to Unlearn. Decolonial Reflections from Eurasia and the Americas*. Columbus: Ohio State University Press.

Tlostanova, Madina, Suruchi Thapar-Björkert and Ina Knobblock 2019. "Do We Need Decolonial Feminism in Sweden?" *NORA - Nordic Journal of Feminist and Gender Research* 27(4): 290–295. DOI: 10.1080/08038740.2019.1641552

Vázquéz, Rolando. 2017. "Precedence, Earth and the Anthropocene: Decolonizing Design", *Design Philosophy Papers* 15(1): 77–91.

Walsh, Catherine and Walter Mignolo. 2018. *On Decoloniality*. Durham: Duke University Press.

Index

Pages in *italics* refer figures and pages followed by n refer notes.

22 de Enero and non-existence of second women 120–121
22 de Enero community 120–122
2015 Oslo Summit on Education for Development 92

Abderrahman, Taha 25
academic decolonialism 147, 159
Adlbi Sibai, Sirin 25, 27–28
Advance Young Chidren's Inclusion and Agency in Society through Voice and Story (ADVOST) 57, 64n1, 64n4
African countries 11
Åhrén, Mattias 72
Alaouf, Leila 23
Ali, Zahra 22–24, 31
Allahu Akbar 32
Allan Poe, Edgar 34
Altamira Square 122
Andean concept of *ayllu* 13
Andersen, Chris 103n19
anti-Muslim racism 37
Anti-Racist Academy in Sweden 5
Arab-Muslim patriarchy 27
Arnott, Lorna 53, 56
art installation, *We are still here* 81, *82*
Asad, Talal 32
Aya, Abdelmumin 26
Azad, Hasan 33

bad faith 9
Badran, Margot 23, 26
Baldwin, James 9
Balto Asta 54
Bañcuţa, Gabriela xii, 3, 126–142
Barlas, Asma 23
Basic Education Act 53, 64n2
Bautista, Rafael 12
Baylan, Ibrahim 75
bedrock, soil and groundwater (SGU) 75, 77, 84n5
Belle, La Vaughn 4
Bentabet, Houssame 33
Bernhardsson, Robert 78
Biccum, April 97
Bimpong, Bright 4
black skins, white masks 9
Blaisdell, Caralyn 53, 56
Bolivarian revolution 111, 113, 119–120, 123
Bolivarian Venezuela 112
Bolivar, Simon 111
Bonilla-Silva, Eduardo 38
Bouteldja, Houria xii, 2, 10, 22–35
British Beowulf Mining 70, 75–78
Building Stronger Universities in Developing Countries 96

capacity for action 37, 47
capitalism 11, 32, 96, 135, 145–146
capitalist hegemony 11
capitalist interests 12
capitalist labour market 9
capitalist-nature ontology 77
Caribbean 2, 111–112, 127
Castro-Gómez, Santiago 149
Chávez Frias, Hugo 111
Chávez government in Chacao 121
Chavista communities 121
Christianity 25, 89, 148
classism 148
Cold War era 146–147, 150
colonial capitalist system 7
colonial complicity 5, 14n4, 71, 76, 154
colonial education 92, 96
colonial forgetfulness/decolonial recollection 70
colonial gender binarism 114

colonialism 26, 97, 127, 148; and coloniality 13
coloniality 1, 5, 73, 112, 114, 145; and Euromodernity 6; form of 154; of knowledge 2–3, 7, 89–104, 155; logic of 89; of power 3, 10, 73, 84, 112, 114
coloniality/modernity 73
the coloniality of being 110, 113
the coloniality of gender 110
colonial settlements 73
colonisation 51, 79, 90, 141, 158
Colonised Arab-Muslim thinking 25
colonization of the mind 9
Committee on the Constitution 75, 84n6
communism 32
Community of Latin American States (CELAC) 112
Copenhagen 43–44, 46
COVID-19 pandemic 90, 93, 100
creolization 140–141; and im/purity 127–128
critical intellectual environments 2
cultural racism 38
cultural relativism/multiculturalism 153
Czarist Empire 158

damnation in Venezuela 110–123
damnés 110, 113, 116–117
Danish: colonialism 94; colonies 1, 4; development 95–96; ethnographic fieldwork among young Muslims 40–43; experiencing Islamophobia 45–46; knowledge and technology 95; Muslims 40, 45, 47–48; national image 44; and non-Western immigrants and descendants 40; political and social system 39; public imaginary 41
decolonial: communities 147, 159; critique 1–2, 148–150, 155; gnosis/border thinking 149; jargon for colonial(ist), hijacking 150–153; to knowledge production and aesthesis 148–150; studies 152; thinking 145–148, 150, 155–156, 158–160; thought 26, 145, 148, 150, 158
decoloniality 145–146, 148, 150, 152, 155, 158–159, 160n1; and decolonisation 13; Europe initiative 160n3; In Europe 152; *for* 7; *from* 7; in Nordic context 153–156; and postcolonialism 12; and poststructuralism 5
Decoloniality Europe network 4
decolonial recollection 74; challenging the virtue of extraction 79–82; concept of 74, 82; locality of extraction 70, *71*; narrative reading of epistemes 69–70
decolonisation 1, 4, 7–13, 55, 63, 147, 151, 155
decolonisation of the mind 74
de-Islamisation 33
de Leeuw, Georgia xii, 3, 68–84
DeliverEd Initiative 92
democracy 7, 27, 37, 48, 71, 89, 93, 111
Denmark 94; *Muslim* minority in 41; Muslims as "Other" in 37, 39; normalised Islamophobia 3
de Sousa Santos, Boaventura 73
digital colonialism 103n17
digital solutions 55, 62
Dominium Maris Balticith project 156, 158
Draăgan, Mihaela 141
duality of self-perception 128
Du Bois, W.E.B. 46
Dussel, Enrique 5, 102n1

education 90–92, 94; gift of 95–99; global education initiatives 91–95; Northern intervention 99; of the South 100
educational intervention 89–104
educational technologies 98
Education Cannot Wait (ECW) 93
Education Commission 90–93, 99–100, 102n7
Education for All 2000–2015 91, 95
Education Outcomes Fund (EOF) 92
Education Workforce Initiative 92
Ehlers, Jeanette 4
El 23 de Enero district 118, 120
empowerment 53, 60–61, 91, 112
epistemic violence 70, 72–74, 76, 82
equality 6–8, 10, 25, 42, 48, 68, 91, 93, 103n16, 104n25, 110–111, 114, 123, 135, 146, 154
Escobar, Arturo 77
ethnonationalism 45
EU citizen residence 129
EU residence permit 138
Eurocentric: epistemic contract 149; knowledge 6; paradigm(s) 2; positionality 149
Eurocentrism 2, 38, 74, 89, 95, 98, 102, 150
Euromodern 150–151, 155; academy 5; conceptual toolbox 148; division 147; endorsement 159; knowledge production system 159

Euromodernity: monopoly 6; and political economy 9
European colonial domination 89
European decolonial trajectory 152
European modernity/rationality 89
European Parliament and European Council 57
European racism 114
European system of knowledge 150
European Union (EU) 57, 110

Facebook community 81, *81*
Fadil, Nadia 39–40
Fanon, Frantz 113
Finland: EU citizen's right of residence in 139–140; Roma women 126; Sami live in 52, *52*
Finnish Aliens Act 136
Finnish gadje women 127
Finnish Immigration Service (Migri) 129, 136–137
Finnish Ministry of Education 64n1
Finnish National Board on Research Integrity 57
foreignness 37
France 2, 23, 29, 31, 33, 35n1, 103n16, 157
Freire, Paulo 13
French constitution 25
French group Lallab 23
French law 35n2
French Republic and individualisation of Muslims: conditional solidarity of white progressives 30–32; individuals and modern states 28–29; surviving like Scheherazade 32–35
French Revolution 25, 33
Fur, Gunlög 156

Gállok 3, 70, 74–76, 80–81, 83
Garner, Steve 46
General Data Protection Regulation (GDPR) 57
Gilroy, Paul 39–40
Glissant, Édouard 129
global education: industry 95; initiatives 91–95; intervention 99; project 90, 95
globalization 26, 29, 91
Global North 38, 159
Global Partnership for Education (GPE) 92; *Raise your Hand—A Case for Investment* 104n22; *Through the Lens of Education* 101
global racial capitalism 11
global secularisation 26
Global South 99, 152, 159
Global Southern perspectives 1, 7
Gordon, Lewis R. 5, 7, 148, 154
Groglopo, Adrián 1–21
Grosfoguel, Ramón 5, 25
guarimbas, racist 121–122
Guerzoni, Michael A. 55
Gypsylorism 126

Hamidi Malika 23
Hassani, Amani xii, 3, 37–48
hegemony 11–13, 34, 112
Helander, Hanna xii, 3, 51–64
Hlabangane, Nokuthula 96, 98
homogeneity 90, 147, 153

Icaza, Rosalba 2
imperialism 6, 11, 27, 145
imperialist politics 2, 11
imperialist wars 8, 11
imperial politics 12
Incheon Declaration for Education 2030 92
Indigenous peoples 3, 51, 54, 57, 63, 84n4, 152, 154, 156
indigenous Sami community 68; *see also* Sami
individualism 22, 29, 34
industrial revolution 5
Ingermanland 158
institutionalisation 27, 155
intellectual extractivism 148
Inter-agency Network for Education in Emergencies (INEE) 93, 103n11
International Finance Facility for Education (IFFEd) 92, 103n14
International Labour Organization's convention No. 169 84n4
International Monetary Fund 103n20
Ipsen, Pernille 156
Irni, Sara 154
Islam: "Din" 35n5; incorporation of 27; political Islam 28, 35n7
Islam, feminist religion: egalitarian religion 24; my body, my choice 23; Quran and Sunna 24
Islamic feminism 2, 22–23, 31–32, 34; in Western modernity 25–28
Islamic feminists 23–24, 26, 32
Islamic reformism 27
Islamism 29
Islamophobes 22
Islamophobia 5, 22, 29; in Denmark 3, 37–47; state racism and 30

Jannok, Sofia 79, 81, *82*
Jensen, Lars 38, 154
Jewish communities in France 33

Kallak 70, 75–78
Kangas, Marjaana 54
Karimi, Hanane 23
Kelkal, Khaled 29
Keskinen, Suvi 154
Keskitalo, Pigga xii, 3, 51–64
Khaled Kelkal in 1995 29, 35n9
Khemilat, Fatima 31
Khiari, Sadri 10, 28
Knobblock, Ina 114, 154
Konstitutionsutskottet (KU) 84n6
Körber, Lill-Ann 72
Kuokkanen, Rauna 51

Lallab Magazine 35n3
Lamrabet, Asma 23, 26
Latin America 2, 4, 8, 111–112, 160n2
Lawrence, Rebecca 72
learning crisis, girl child and South 91; education in the South 100; global education intervention 99
León, Marilin 119
Lindstad, Gro 103n20
Linnaeus University Centre for Concurrences in Colonial and Postcolonial Studies 153
Loftsdottir, Kristín 154
Löfven, Stefan 68, 76
Look Who's Talking: Eliciting the Voice of Children from Birth to Seven 56
Lubienski, Christopher 102n3
Lugones, Maria 114, 128–129

Mahmood, Saba 47
Maldonado-Torres, Nelson 73, 112–113, 117, 122
Märak, Maxida 80
Märak, Mimi 80–81
Marxist critique of capitalism 146
masculinities 27, 31
Matamoros, Zuleika 118–120
Mechai, Hassina 33
Mignolo, Walter 9, 73, 89, 147, 149
Mining for generations—Mining by Sweden 76
mining industry 68, 70, 75–76, 78, 81, 84n8
Mir-Hosseini Ziba 23–24, 27
modern capitalist economy 9
modern (European) civilization 90
modernisation 89, 94, 129
Moorhead, Shelley 4
Moosavi, Leon 151
Mörkenstam, Ulf 72
Morrison, Toni 103n18
mosque 43–45, 48
Mulinari, Diana 154
Mun biepmu dutnje attán 61
Muslim(s): agency 37, 46–47; civic Islam 29; community 11, 22–23, 32–33; in Copenhagen 37, 43–44; jurists 34; masculinity 34; men 27, 41–42; Other 26, 37–40, 42, 44–45, 48; Otherisation of minorities 47; *Otherise* 46; patriarchy 27, 31, 34; racialization of 46; *see also* racialisation in Denmark; as self-ascribed and racialised 43–45; societies 24, 27, 29, 43; subject and everyday life 46–47; threat 37; vision of liberal Islam 30, 33; women 22, 26–27, 29–30, 42, 45, 48; youths 3, 5, 38
Muslim Brotherhood in Egypt 31
Muslimness 10, 37, 40–41, 46
Myrdal, Gunnar and Alva 8
myth of Danish leadership 94
myth of modernity 5–6, 12

National Autonomous University of Mexico (UNAM) 34, 35n14
National Constituent Assembly 120
National Electoral Council (CNE) 120
nationalism 27–28, 39–40, 51
national self-imagery 38
NATO-Russia war in Ukraine 12
Navarro, Tami 4
Ndlovu-Gatsheni, Sabelo J. 92, 151
neoliberalism 5, 27, 29, 112
neoliberal politics 11
Netherlands 39
News Network *TeleSur* 112
the Night of Culture 43–44
Norborg, Bengt 115–118
Nordic academia 5, 11
Nordic actors 89, 93
Nordic compensation 158
Nordic countries 4, 8, 10, 12, 93, 101–102, 113–114, 154, 156
Nordic education 89–104
Nordic Europe 4, 145
Nordic feminists 10
Nordic imperial-colonial configuration and trajectory 156
Nordic region 4, 158; areas of study in 2; coloniality and decolonisation in 7–13; experience of people in 1; investment 11–12; Muslims 10; postcolonial and the decolonial 4–7

Nordic settler and internal colonialism 156
Nordic societies 5, 7, 10
Nordic style 156–158
Nordic universities 2
Nordic welfare state 154
Nordin, Jonas 69
Nørrebro 44
Norrland 79–82
North Atlantic blockade 123
North Atlantic corporate media (CNN 2017) 112, 119, 122
North Atlantic neoliberalism 112
North Atlantic powers 115–116
North Atlantic Treaty Organization (NATO) 11–12
Northern Sami 52
Norway 8, 14n1, 52, 92, 94, 101, 102n5, 103n16, 104n23–104n24
Norwegian international 93
Norwegian Ministry of Foreign Affairs 103n15
Norwegian *Oil for Development* programme 97
Novgorod Republic 158

Office of the United Nations High Commissioner for Refugees (UNHCR) 93, 103n10
official Norwegian aid statistics 93–94
Ojala, Carl-Gösta 69
Øland, Trine 8
Omi, Michael 38
One Thousand and One Nights 34
Organization of American States (OAE) 112
orientalism 126, 152
Other(s) 26, 37–40, 42, 44–45, 48, 71, 73–74, 94, 100, 102n2, 126
the Other 73
Otherisation 8, 47
Otherise 46

Padovan-Özdemir, Marta 8
Pasqua, Charles 29
pedagogical innovation 55–57, 62
pedagogy of cruelty 110
PetroCaribe 112
Pieski, Satu-Marjut xii, 3, 51–64
Pilot project 53–54, 62
playful learning: approach and purpose 62–63; define and listen 57–58; inclusion and empowerment 60–61; process and structure 58–60; as resistance 54–55
political nationalism 40
politics of purity 128
positionality 5
postcolonial 153; Anglophone and Francophone 156
postcolonial theory 2, 145, 151–152, 155
poststructuralism 5–6, 148, 154
privileges 8, 32, 74, 132, 141
progressive humanitarians 8
progressivism 23, 30, 32, 151

Qatar 43
Quijano, Aníbal 38, 89, 112, 146
Quran: patriarchal interpretation of 26; and Sunna 24
Quranic decorative scripture 43

race and racism 8, 39, 98, 148; and colonization 98
racial capitalism 1–2, 4, 6, 8, 158
racialisation in Denmark 37; civilised nation and "threatening" Other 39–40; in nation without race 38–39
racial taxonomising 145
Randolph, Justus 54
real Danes 45
rebel group *tjáhppis rájddo* 80
reductionism 102
Reinfeldt, Fredrik 68
Renberg, Elsa Laula 82
Results in Education for All Children (REACH) 103n13
Robinson, Carol 53, 56
Robinson, Jenny Perlman 55
Roma communities 135
Roma Futurism 141
Roman culture, storytelling 128–130
Romania 133
Romanian and Bulgarian Roma: migrants 126; women 127
Romanian gadje women 126–127, 141n2
Romani literature 129
Roma people 128, 141
Roosevelt corollary 111
Ruokamo, Heli 54
Russian empire 157
Russian Federation 52, 157

Said, Edward W. 101
Sami 158; animals 61; and environmental opposition 70; flag gazing over land 79; live in Finland 52; population 77; resistance in Gállok 3; tradition and culture 80–81; Utsjoki, Inari, Sodankylä (home land) 52

Sami children 3, 51, 61
Sami community 68, 75
Sami culture 54–55, 61, 63
Sami education 51, 53, 56
Sami homeland 53, 58, *71*
Sami language distance education 51, 63, 64n1; expansion 54; pedagogical innovation and analytical framework 55–57; research questions 53; toyshop play 58–60, *59*
Sami people 72, 80; colonisation of 51; in Norway, Sweden, Finland and Russia 52, *52*; and violence 83
Sápmi 68–69, 78, 80, 82; colouring 79; Gállok in Julevsápmi (Lule Sami) *71*; land song *This is my land* 79; as Swedish America and Land of the Future 72
Save Our Future campaign 91
Scandinavia 14n1, 72, 156
Scandinavian: colonialism 156; countries 104n23, 157–158
Segato, Rita 114–115
Selod, Saher 46
SIDA's education portfolio 104
Siklander, Pirkko 54
Sinclair-Poulton, Clive 75, 81
Smith, Linda Tuhiwai 103n19
South and Indigenous communities 9
Southern Other 94, 100
South/Global South 102n1
Soviet Union 157
spiritual brain-drain 9
Spivak, Gayatri 118, 154
Steiner-Khamsi, Gita 102n3
Støjberg, Inger 42
Stormaktstiden 156
Structural adjustment programmes 99, 103n20
Suárez-Krabbe, Julia 1-21, 96
suma qamaña, Kichua concept 13
Sundström, Bert 122
Sunna, Anders 81, *82*
Sustainable Development Goals (SDGs) 92
SVT reporting on Venezuela: coloniality of gender and pedagogy of cruelty 113–115; experience of damnation in Chacao 122–123; second women 119–120; *On the Verge of Ruin* 115–119
Sweden 76, 156–157; colonial neutrality 71; mining identity 68; visions of the North 77
Swedish: colonial exceptionalism 154; colonialism 69, 72, 81–82; complicity 157; extractive exceptionalism 82–84; government 76; and Nordic colonial past 72; resource abundance 76
Swedish Association of Mines, Minerals and Metal Producers 76
Swedish exceptionalism 69–70, 79; colonial/extractive episteme 72–74, 84; race biology 71; Sami people 72
The Swedish Mine 79
Swedish mining 78, 83; companies marketing campaign 79, *79*; as exceptional 76; and Gállok/Kallak 75–76; image of 77; industry 70; policies 72; story of 76
Swedish Public Television Service (SVT): colonial ideas of gender equality 111; production of non-existence of second women 120–121; Venezuelan crisis 110; *On the Verge of Ruin* 110; website 110
Swedish tourism industry 80
Swedish virtue of mining 73

Tafari-Ama, Imani 4
Tamatea, Laurence 95
Thapar-Björkertb, S. 114, 154
Thiong'o, wa Ngũgĩ 74
Third World 94, 101
Third World Girl 101
Third World Woman 100–101
Thorwaldsson, Karl-Petter 70, 75
Ţîştea, Ioana xii–xiii, 3, 126–142
Tjáhppis Rájddo manifestation *80*
Tlostanova, Madina xiii, 3–4, 114, 145–160
Torres, Rosa María 91–92
toyshop play 58–60, *59*
traditional Sami child-rearing 54
Trump, Donald 118
Tuori, Salla 154

UN Committee on the Elimination of Racial Discrimination (CERD) 75
undertheoretisation 5, 7
United Nations Children's Fund (UNICEF) 93, 103n9–103n10
United Nations Development Project (UNDP) 93, 103n10
United Nations Educational, Scientific and Cultural Organization (UNESCO) 75, 91, 93, 97
United Socialist Party of Venezuela (PSUV) 118

United States of America (US) 8, 35, 45, 110, 117, 119, 160n2; US imperialism 111–112
Universal Declaration of Human Rights 91
University of Gothenburg 14n3
University of Lapland 57

Vázquez, Rolando 2
Velásquez Atehortúa, Juan xiii, 3, 110–123
Venezuela: as historical threat to empire 111–112; naturalisation of war and the damnés 112–113; SVT reporting *damnation* in *see* SVT reporting on Venezuela
Venezuelan government 113
Verger, Anthony 102n3
Vićentić, Jelena xiii, 3, 10, 89-109

Wahhabis in Qatar 43
Wall, Kate 53, 56
Walsh, Catherine 7, 89
Walter, Maggie 103n19
War on Terror 37
Wekker, Gloria 39–40
Western-centric paradigm 23
Western civilisation 37
Western epistemology 149
Western feminism 2, 25
Western freedom 27
Western hegemony 28
Western heterosexism 27
Western imperialism 9
Westernised universities 6
Western knowledge 94, 96
Western modernity 25–26, 73
Western societies 25
Whilst Orientalism 126
white Danes 48
white feminism 10, 26
White Innocence 39
white mask 9–10
White Paper 25 Education for Development (2013–2014) 97
Whites, Jews and Us: Toward a politics of revolutionary love 32
white supremacy 6, 8–9, 11, 31, 132, 141n2
Winant, Howard 38
women liberation 23
World Bank 93, 98, 101, 103n13, 103n20
World Declaration on Education for All (EFA) 91–92
world economy 96
world-travelling stories: *Able to defend myself* 137–139; *Dream big* 139; *keep walking* 136–137; *'Liberation'?* 139–140; *loud silence* 130–132; *A morning swim* 132–133; *My mother makes the best polenta in the world 134*, 134–136; *The pan in the system* 133

Yanique, Tiphanie 4
Yasin, Nadia 28
young Muslims 40–43

For Product Safety Concerns and Information please contact our EU
representative GPSR@taylorandfrancis.com
Taylor & Francis Verlag GmbH, Kaufingerstraße 24, 80331 München, Germany

www.ingramcontent.com/pod-product-compliance
Ingram Content Group UK Ltd.
Pitfield, Milton Keynes, MK11 3LW, UK
UKHW021307260926
475702UK00048B/804

9781032274867